Closed-circuit
Television
Single-handed

Closed-circuit Television Single-handed

TONY GIBSON MA

*Head of the Audio-visual Centre and the
Television Research and Training Unit,
University of London, Goldsmiths' College*

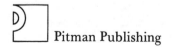

Pitman Publishing

First published 1972

Sir Isaac Pitman and Sons Ltd
Pitman House, Parker Street, Kingsway, London WC2B 5PB
PO Box 46038, Portal Street, Nairobi, Kenya

Sir Isaac Pitman (Aust) Pty Ltd
Pitman House, 158 Bouverie Street, Carlton, Victoria 3053, Australia

Pitman Publishing Company SA Ltd
PO Box No 11231, Johannesburg, South Africa

Pitman Publishing Corporation
6 East 43rd Street, New York, NY 10017, USA

Sir Isaac Pitman (Canada) Ltd
495 Wellington Street West, Toronto 135, Canada

The Copp Clark Publishing Company
517 Wellington Street West, Toronto 135, Canada

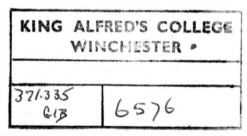
ISBN: 0 273 43930 8

Text set in 11/12 pt. IBM Baskerville, printed by photolithography,
and bound in Great Britain at The Pitman Press, Bath
G2.G3349:15

Acknowledgements

My colleagues in the Television Research and Training Unit at Goldsmiths' have been as patient and considerate as ever. Besides figuring in some of the illustrations, they have helped to devise or develop many of the techniques demonstrated. To them as a team —Paul Barnes, Brian Bridger, Hazel Fullerton, John Hughes, Jo' Peters, Stephen Woolhouse—this book is respectfully dedicated. Professor Roger Warwick and Mr A. N. Finch gave especial help in demonstrating their television work in the Department of Anatomy, Guy's Hospital Medical School.

Donald Nisbet took most of the photographs. Additional illustrations were by courtesy of Bell & Howell Video Systems; General Video Systems Ltd; Jo' Peters; Alan Thomson; UNESCO; and the School of Art, Watford College of Technology. Teltron Ltd now manufacture the Goldsmiths' multipurpose studio illustrated in pictures 47, 53, 103, 117 and 121–5.

David McClelland, director of Television, NE London Polytechnic, devised the chart on pages 137–9.

T.G.

Contents

I

Wide-angle View

This book is dedicated to the proposition that intelligent users of closed-circuit television need to know how to walk before they try to run but that they do not want to go on walking for ever.

It assumes that, to begin with, many users must be ready to fend largely for themselves, operating equipment single-handed in a school, a hospital, a factory, a laboratory, a management centre, where there will sometimes at best be only part-time help from a qualified technician for overhaul and repairs. This means learning how to exploit limited resources imaginatively, knowing what each item of equipment is capable of doing and then concentrating on the techniques which yield the best results with the least fuss.

Besides showing what two or more television cameras can do, it will also help the user to understand what one camera can sometimes do even better than two, and how television can gain by association with film, slides and audio-tapes.

Effective use depends on suiting the hardware to the needs of the job in hand, rather than trying to reproduce a form of television production that was evolved to meet quite different needs and depends on much more elaborate resources. This does not mean that the traditional type of broadcast programme is beyond the scope of closed-circuit television; but it is a great deal less relevant when you consider the special advantages that even the simplest type of closed-circuit system can offer. These advantages can be seized even by the beginner, if he thinks for himself and deploys his equipment with common sense. Once aware of the possibilities he will find himself developing his own techniques to

make more imaginative use of his resources, to solve bigger problems and to grasp fresh opportunities with greater certainty and
less fuss.

Meanwhile, the first step is to see just where you are going.

Special Characteristics of Closed-circuit Television

The problem in any technology is to see the wood in spite of the
trees. What follows is intended for those who, besides being
foresters and woodcutters, must know the uses of the timber they
plant and harvest. Closed-circuit television has more applications
than at first meet the eye. It is a building material, to construct the
furniture of the mind, explaining processes and concepts; it is a
raft on which to convey the raw material of everyday experience;
it is tinder to fire curiosity, kindle sympathies, inflame imagination.

So the enlightened user will see himself as cabinet-maker, boat-
builder, fire-maker, exploiting the properties of the material for
many different applications, and in many different situations.

The variety of situations closed-circuit television makes possible
is the key to understanding the special advantages that closed-
circuit possesses over network broadcasting.

Consider the five elements in a television situation: subject
matter, camera(s), programme-maker, screen(s), and viewers.

The *subject* could be: a surgeon at work in an operating theatre,
or an anatomist in the autopsy room; a personnel officer
demonstrating interview techniques; the central control console in
a power station; a University lecturer in full spate; an architect's
townscape, seen from the viewpoint a townsman might have at
street level; a production line on the shop floor; a tiny specimen of
pond life; a piece of flawed metal in a metallurgical microscope; a
group discussion; a role-play, or a dramatic interlude; a cookery
demonstration; the hands of a musician at the keyboard; a traffic
snarl-up; a teach-in; an evocative sequence of film, slides and still
photographs; an abstract composition of changing light and shade;
a mathematical sequence on blackboard or computer tape; children,
or birds or insects, or reptiles seen close to, undisturbed, as they
normally are.

The *camera(s)* might be remotely controlled; or manipulated b'
an operator at the behest of a director far away; or in a fixed pos'
tion where they command a process or an activity; or hand-held '
the programme-maker himself, moving to follow the action
wherever it goes.

The *programme-maker* might be a television specialist, deploying all the resources of film animation, dramatization, eye-to-eye contact; he might be a research worker using a television camera just as he might any other laboratory tool; or he might be no more than a copier of somebody else's programme, recording it off the air, or off a cable network from a local authority system.

The *screen* may be one of many distributed throughout a building, or across a university campus, or at different strategic points in an industrial process. Or just one screen may be sufficient—for the investigator using a tiny camera to probe the recesses of a long pipeline to establish where corrosion is under way, or for the gymnast observing himself as he practises a crucial movement. It may reveal what is so dangerously or awkwardly placed that no human eye can observe it, or it may simply supplement what the viewer can himself see direct, but not to such great advantage.

As for the *viewers*, they may be all together in one place, and at the same time; or scattered in many different rooms; or watching at many different times as material that was seen live on the first occasion is replayed from a video-tape-recorder as often as the need arises.

So much for the ingredients. Now compare the different ways in which they can be combined (1, page 4).

The permutations rival anything one might encounter on the Pools! Perhaps this bewildering variety explains the slight misunderstanding that arose when a group of eminent educationists came together for a demonstration of the possibilities of closed-circuit television. At the end of an hour they had been shown practically every trick in the locker. They had been persuaded to demonstrate craft skills, to conduct bench experiments, to manipulate materials, to perform simple animations, to interview each other, and to talk to camera. Throughout they kept watch on the effects of these operations, live, on the television screen.

At the end of it all one visitor button-holed the camera operator and politely asked if he could be shown "where you put that cine film in"!

Closed-circuit television takes a bit of getting used to. We have so many preconceived notions. We may confuse it with other media, such as film and slide; or judge its possibilities and limitations entirely in terms of the broadcast material we are used to seeing on the screen back home.

There is plenty in common with broadcasting, of course, but

1

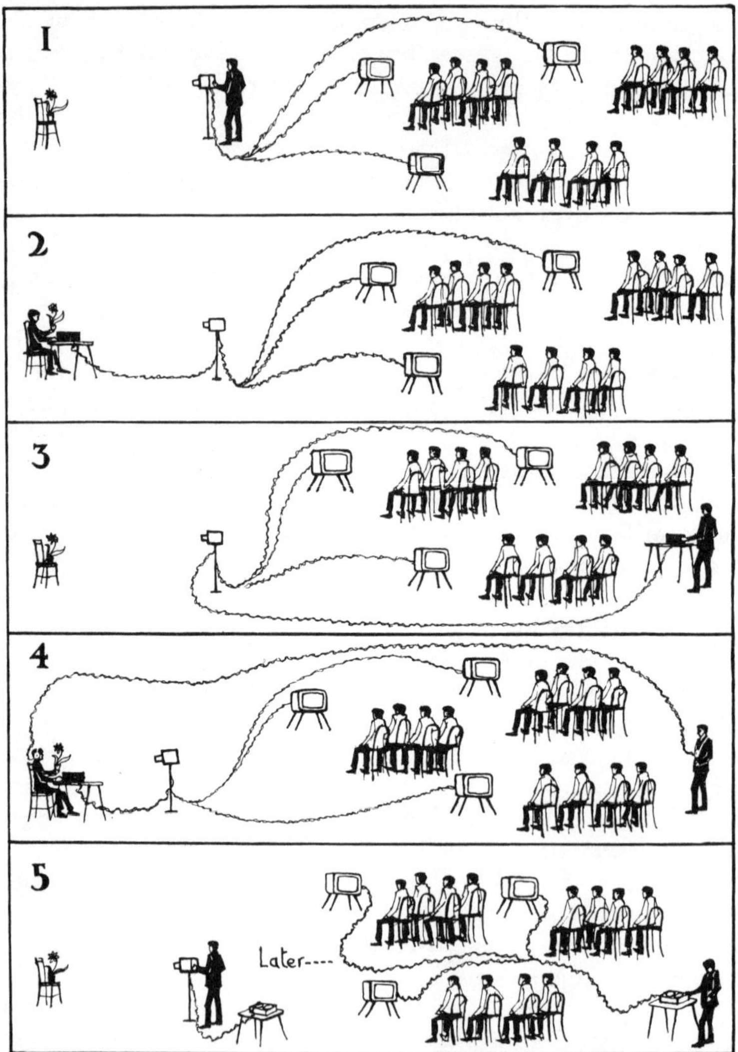

closed-circuit television cuts out the detour which the signal fr$
the broadcasting studio has to make before it is picked up on t!
domestic receiver. Everything in closed-circuit television is on t
same closed line. It is within reach, and can be put under contr
We can keep track of the process from start to finish, and gradu
ally recognize the job each piece of equipment performs.

How the Image Reaches the Screen[1]

The *lens* of the television camera does almost exactly the same job
as the lens of a photographic camera, and provided that the dia-
meter of the lens mount and its screw fitting is the same a film-
camera lens can be used on a television camera in emergency.

In television, as in photography, the image of the subject is
projected by the lens on to a piece of light-sensitive material. In
photography this is a film or photographic plate. In television the
light-sensitive area is called the *target* and it forms part of the
television tube. The *tube* is the most fragile component of the
television camera, and it consists of an elongated glass bulb within
which a narrow beam is manipulated so that it traverses the target
area in much the same way as lines of typing traverse a sheet of
paper. This process is called *scanning*; the beam moves across each
horizontal line from left to right (*line*-scanning), and then,
instantaneously, resumes below from left to right once again. This
gradual progression from top to bottom is called *frame*-scanning.
When the beam gets to the foot of the page, so to speak, it starts
anew at the top left-hand corner. The special quality of this beam
is that it produces an electrical signal that is varied by the light
intensity it encounters on the target. One might compare these
variations with the separate letters or spaces in a line of typing. At
each point the beam registers the darkness or lightness of the image
formed by the lens at that particular point.

These variations are reproduced at the viewing end by the beam
which operates on the inside of a cathode ray tube in the television
screen. The scanning movements of the beams are synchronized by
a pulse generator. The whole scanning process is so fast that it
deceives the human eye. What is actually a dot of light that passes
ack and forth across the television screen, waxing and waning as
: encounters dark or light parts of the picture, appears to us as a
ontinuous trail of light, as we see it so fast that we do not recog-
ize that the picture on our screen is really a single track, line

This relates particularly to the vidicon type of camera tube and system.

blending with line to create the impression of a continuous picture area. The deception is helped by the fact that the beam traverses the odd lines first, and on the next run, "down the page," it traverses the even lines. The effect is to *interlace* the two patterns and so disguise the grid effect. A *fixed* interlace ensures that the lines never duplicate each other. In less sophisticated cameras the patterns do sometimes overlap, and the system is called a *random* interlace. When more than one camera is in use on the same assignment more precision is required to get each camera in step with the others. This can be achieved by making one camera drive the others (a *master-slave* system), but this may not be reliable for more than three cameras at a time, and a preferred system is to set the pace for all the cameras by introducing a *synchronizing pulse generator* (s.p.g.). This controls *line drive* (L.D.), *field drive* (F.D.), and *mixed syncs.* (M.S.)—the odd and even line patterns.

The screens used by those who operate the television equipment are called *monitors*, and they may be placed independently in studio, or control cubicle, or on location (to help the demonstrator see what he is doing); or small screens, known as *viewfinder monitors*, may be attached to individual cameras.

If more than one camera is being used to cover the same subject, a *vision-switcher* is required to allow for alternation between one camera's picture and the next. If both are to be relayed to the viewer at the same time (when the pictures are blended in a mix, or an inlay, or a superimposition) a *vision-mixer* is required, and this usually incorporates the vision-switcher.

The pictures formed by the cameras are linked with a separate sound system, and fed by cables to the screens the viewers see.

When it is desired to record what is televised, another link is inserted in the chain. The changing patterns of electrical impulses which television camera(s) and sound system provide are relayed to a *video-tape-recorder* which translates them into a magnetic pattern, imprinted on the tape as it passes the *recording head* of the machine. To reproduce sound and vision together afterwards, the tape is spooled back again and replayed so that when it passes a *replay head*, the magnetic patterns are re-translated into patterns of electrical impulses; these in turn are transformed into variations of light intensity and sound intensity on the television screen and its loudspeaker.

2

Hand and Eye

Getting Sound and Vision Right

Responsibility begins with the engineers who design the equipment, manufacture it and service it. Different items may well have been manufactured by different firms and require careful matching-up at the time they are purchased. This matching-up is checked over at the commissioning stage, and no item should be paid for until commissioning has been successfully completed on the spot where the equipment will be in use. This is the moment to make sure that along with each of the major items of equipment the manufacturer supplies the circuit diagrams and the manual of operation and maintenance.

The responsibilities now begin to mount on the shoulders of the operator himself. He has to keep an eye on the performance of the equipment, to note whether one item intermittently seems faulty, and may need replacement or repair; and to make sure that when equipment is stored away, or packed up for transportation, it is kept from harm by dampness, heat, excessive jolting, or the inquisitive fingers of the passer-by. Equipment that has been properly designed will not be very vulnerable to these hazards. It will be light in weight, easy to carry and to reassemble, with robust and simple connections and easily identified controls. Above all, it will be versatile, so that the same equipment can be rearranged in different formats, to suit different requirements and situations.

The operator's first job is to familiarize himself with this range of possibilities, so that he gets thoroughly used to putting the

:quipment together in the way that best suits a particular assign-
nent and wastes least time and space.

The next thing is to get clear in one's mind the dividing line
)etween what an engineer should be asked to see to, and what re-
nains for the operator to check over, manipulate and maintain.

The *lens controls*, which form an accurate image of the
subject, in sharp focus, on the target area, should be his concern
entirely as regards cleaning, occasional lubricating, and operating.

The positioning or *racking* of the tube in relation to the lens,
which makes it possible to lengthen or shorten the distance
between the target area and the back of the lens, so as to cope
with subjects viewed so close that they would otherwise be out of
focus, should be the concern of the engineer, unless the camera is
constructed with an additional facility for the operator to manipulate the racking device whilst he is operating his camera. On the
more sophisticated cameras this makes it possible for the operator
to cover subjects a few inches away from his camera, or a few
feet, without having to stop to screw on a supplementary lens for
very big close-ups. Otherwise the rack distance is best pre-set for
normal use, which means use within the distances for which the
lens itself was designed.

The range of light intensity relayed by the beam is regulated by
"clamping down" on either extreme so as to allow the beam to
recover quickly from registering one end of the light scale. Correct
adjustment is part of the normal servicing of the camera by the
engineer.

The *pulse generator* determines the frequency and the regularity
of the "beat" by which the beam distinguishes each point along
the line it scans. In simple equipment it is part of the camera. In a
slightly more ambitious set-up a sync. pulse generator is involved.
In either case correction should be the responsibility of the
engineer.

The *raster* determines how large shall be the target area on the
tube. If for example the raster is too small, some of the outer
margins of the image projected by the lens will not be relayed. On
the simpler types of camera there may be a tendency for the
picture to *drift*, and the operator should make his daily check
when he lines up his cameras and correct this if necessary.

The light-sensitivity of the *tube* on to which this image is projected will vary according to the type of tube used. Note that
some types of vidicon tube are more vulnerable to damage by the

sudden concentration of excessive light, which may leave a scar on the tube which blemishes the picture for hours or for weeks. Old or damaged tubes need replacement, and the need for this is established by logging the amount of use, and by checking picture quality each day, using a test card.

The *vision-mixer*, if properly designed, should give little trouble, but the faders, which are linked to the vision-mixing lever, can accumulate dust, and as a result affect the picture which is relayed to the viewer's screen. This can be misleading, and the operator will blame the monitor, or the cameras, unless he carefully compares the unblemished viewfinder monitors with the faulty picture on the screen fed by the mixer and so locates the trouble.

The *monitors*, and the screens used by the viewing audience, need careful adjustment, just as the ordinary domestic television screen does. And in just the same way the troubles usually arise because somebody has been fiddling with brightness, contrast, vertical and horizontal line holds, without doing so systematically. Turn up the contrast and brilliance controls almost fully, i.e. clockwise. A pattern of lines should appear on the screen. Retard the brilliance control until the lines just disappear. Retard the contrast until the picture appears in its natural tones. If it is unstable, try vertical and horizontal line holds, one at a time.

Discourage any further manipulation of screen controls, because from now on the screens will be the guide to the effective *adjustment of the camera controls*. These are beam and target adjustments, which on some cameras will have their own separate knobs, and on others will require manipulation with a screwdriver. Once the appropriate screws have been located this is a straightforward proceeding which the operator should be prepared to tackle under the guidance of his engineer. A weekly check will usually be enough. Aim the camera, with lens at maximum aperture, at the lightest subject it will be required to view. Begin by turning both beam and target down to zero. Turn up the beam until a picture emerges. Turn up the target to the minimum to achieve a good contrast. Readjust the beam to get the white areas as white as possible.

With this background information in mind we can now proceed.

The Daily Line-up

Arrange the basic equipment to suit the day's requirements. This includes the placing of monitors where demonstrators who may have to refer to them can see them from the positions they will occupy. Check all power plugs and cable connections, to make sure there is nothing loose that could cause an intermittent fault. Coil up unwanted cable and arrange what is left so that it is out of people's way and looks inconspicuous. Where it is in danger from people's feet cover it with rubber mats or wooden duckboards. Remove all unwanted clutter.

Inspect lens and tube for dirt, and clean as necessary. Use lens tissues or a chamois leather to clean the tube (the surface reached through the screw mount), and both ends of the lens (which when not in use should be kept safely in the box provided). Keep the lens capped when not in use. (2, 3.)

2

3

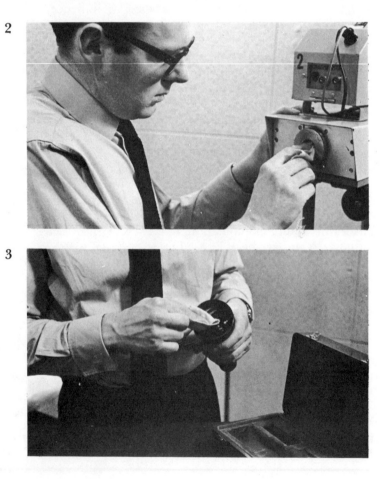

Set up duplicate test cards in front of the cameras, each one receiving the same illumination evenly distributed. Switch out all other lights, apart from a general working light. Check the distances (camera to chart and lamp to chart), so that they are the same for each camera. Use the same type of lens on each camera, at the same adjustment of focus, zoom and aperture.

4

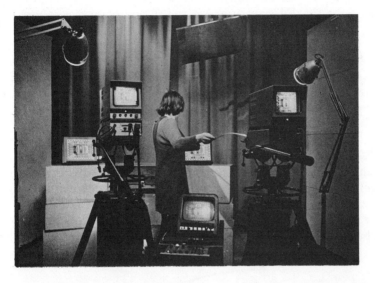

Adjust the viewfinder monitors and the programme monitor (the screen that shows the picture, or mixture of pictures, the viewers will see) so that the rasters (conformation of image to screen) correspond. When the cameras are correctly matched, it should be possible to mix from one camera to the other without any change appearing in the test-card image on the programme monitor. If the images do not match adjust one camera at a time, observing the effect on the programme monitor. Make final adjustments to brightness and contrast, first on the programme monitor, afterwards on each of the viewfinder monitors.

Now replace the test cards by caption cards of plain grey. Any blemishes, or dirt, on the camera tubes should now be revealed. The dirt can probably be removed by fresh attention to lens and tube; the blemishes may be scars caused by over-exposure to bright patches of light, and these may take hours or days to fade. They may persist, and in that case the only remedy is replacement. Even if you have to go on using a scarred tube, it helps to know which is the better of the cameras available and to reserve the better camera for the more exacting shots.

Check the movements on the pan/tilt heads, and lubricate sparingly where necessary.

Adjust the angle and length of the pan handle on each camera to suit your operation. This may not mean identical adjustment. In picture 4, pan handles are arranged to suit an operator who will manage both cameras and the vision-mixer. The length and angle of the pan handle relate partly to the operator's height, and partly to the subject tackled. Slow traversing of the subject matter in vision may be helped by extending the pan handle to its full length in order to increase the leverage. Much quick manipulation, re-aligning cameras between shots, might be done better with a shorter handle.

Arrangement of Equipment

This should be both simple and compact, to permit easy access by the demonstrator to each display area and unfettered movement for the operator.

The diagram shows how the two cameras and a vision-mixer have been designed as one unit, so that a single operator can tackle

5

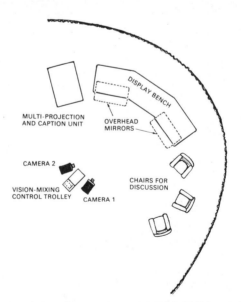

MULTI-PROJECTION
AND CAPTION UNIT

DISPLAY BENCH

OVERHEAD
MIRRORS

CAMERA 2

CHAIRS FOR
DISCUSSION

VISION-MIXING
CONTROL TROLLEY

CAMERA 1

all three jobs without outside help. Display material is arranged on
arcs, so that, as far as possible, each item is equidistant from the
camera which commands it. (Camera 1 is at the hub of a horizontal
arc, Camera 2 at the hub of an arc in the vertical plane, using
mirrors. See also 71.) This greatly reduces the need for adjustment
of focus. The cameras are placed as close together as free manipulation permits, so that the operator has everything she requires
within easy reach. Changes in viewpoints are achieved by the use
of mirrors, described on pages 48–59.

But you cannot always choose your ground. Here the operator
is fitting his equipment into the space between two laboratory
benches (for the lecture-demonstration shown on pages 125–6).

6

The right-hand bench is in use by students. The left-hand bench
supports the programme monitor to which the lecturer will refer
when drawing diagrams on the frame which the operator is setting
in position. He checks everything he does by reference to the
programme monitor, taking up the demonstrator's viewpoint, and
arranging equipment so that it suits the convenience of the
demonstrator as well as meeting the requirements of the cameras.
The vision-mixer here is placed between the two benches (left-hand
corner of picture) so that once the demonstration begins the

operator can stand clear of the cameras (which for this demonstration will need little subsequent adjustment). Compare the pictures on page 125 with the studio arrangement of the same equipment covering a similar assignment, on pages 67—8.

Camera Safety

Safe operation of cameras requires attention to the pan/tilt head. If the camera is top-heavy make sure that the tilt movement is locked after adjustment; there is then no risk, when you let go of the pan handle, that the camera will tilt abruptly forward and perhaps damage the lens on a projecting working surface.

At all costs avoid exposing the vidicon tube to a damaging concentration of bright light. The American astronauts twice made this mistake, on Apollo Mission 12 when they got too much reflected light from the Earth's surface, and on Mission 14 when their camera tube was irreparably damaged as the result of exposure ⟩ the sun.

Your own hazards may be less spectacular but just as exasperatɪg. Inadvertently a camera may be panned, or zoomed out, so hat it unexpectedly includes a source of bright light—a lamp used o accent the subject matter, or just possibly the *image* of a lamp in ι mirror that has been pushed out of its correct position.

Prolonged exposure to a pattern of bright contrasts is also dangerous. If, during a discussion or a tea-break, a camera is left aligned on a sharply contrasted caption, or on a shot which includes a patch of bright sunlight or of bright sky seen out of the window, the pattern will gradually be etched on to the vidicon tube, and will take time to fade off or will permanently scar it. So, whenever there is a break in rehearsal or production, shut down the aperture, and defocus. If the subject is brightly lit and boldly patterned, pan the camera off it on to a less contrasty background.

Manipulation

More sophisticated zoom lenses (such as this Angénieux 1:10 in 7) provide two methods of zoom adjustment. A small handle is being rotated to give a very gradual zoom effect. When this is not required, the capstan handle is pressed in towards the lens, and this dislocates the mechanism.

The bar shown sticking out from the upper surface of the lens enables the operator to adjust the zoom quickly. Some operators prefer to rely on this even for slow zooms in vision.

7

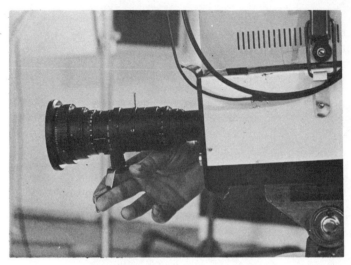

8 and 9 show ways of manipulating camera and lens to get the right shot. In 8 simple cameras, without viewfinder monitors, are manipulated by an operator using a programme monitor he shares with the demonstrator. (Compare 13 on page 19.) Each camera has a 1:4 Angénieux zoom lens, fitted with a bar for zoom manipulation. The right-hand camera has a lens turret, so that other, fixed-focus, lenses may be used. Here, a 150 mm lens affords very close observation of small specimens placed nearby. A 25 mm fixed-focus lens offers a wider-angled view of activities in a cramped classroom or workshop than a zoom lens can command.

8

In 9 the operator must stand back a little to make use of the viewfinder monitor, but she keeps as close to the cameras as she can in order to control its movement by moving her body with it when she pans.

9

Locking devices vary from one type of camera mount to another. In 4 the tilt lock is controlled by the bar (with knobs on either end) shown in silhouette against the left-hand caption and the control for locking the panning movement by a similar bar on the right of the pan/tilt mount. In the more primitive mounts shown in 8 and 9 the tilting movement is locked by rotating the pan handle; the horizontal movement may also be locked, by a screw on the surface of the mount, but this should be used with great discretion. It is easy to forget that you have locked it, and to initiate a panning movement which gradually unscrews the pan/tilt mount from the head of the shaft.

10

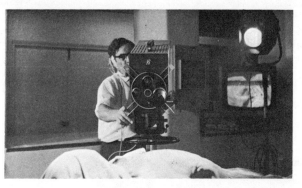

The lens turret shown is used with a more sophisticated camera, equipped with its own viewfinder monitor. The turret can be operated by the "windmill" levers, each marked with a distinguishing colour to facilitate accurate lens changes. The larger the camera, the longer must be the operator's reach (unless the camera pedestal permits sideways operation as in 47). Some users rely upon a remotely controlled zoom lens in which Bowden cables substitute for the long arm of the operator, so that he can get right behind the viewfinder monitor and see what he is doing when he adjusts a lens position. As even sophisticated cameras become smaller and more easily manipulated, remote zoom control, because of its additional weight, becomes more of a liability and less of an asset.

The Operator's Eye-line
In 11 and 12 each operator has a camera with its own viewfinder monitor, which neither is using at this moment. Why?

1

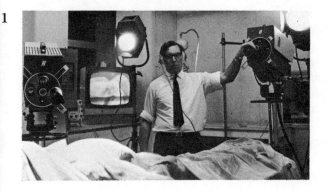

The operator must divide attention between what the camera is showing and what it might be required to show. In this operating theatre there is an additional programme monitor which shows him what the viewers will see, but he needs also to be on the watch for moves by the surgeon.

In 12 the operator is preparing to *soar and swoop*. The demonstrator is sometimes showing the material on his bench to the camera, and talking about it; and sometimes talking direct to

12

the camera, to make a general point without the distraction of his bench material. The operator can use one and the same camera to secure three shots in succession: close-up of the material; zooming out for a medium shot in which the material and the demonstrator's face are included; swooping in again, this time on the demonstrator's face for a close-up. It is important to include this intermediate "soaring" shot if the sequence is to be done by a single camera. If the lens remains zoomed in for close-up throughout there is a clumsy transition as the camera slithers upwards from material to face.

The operator has her eye on the demonstrator to see what he

will do next, although her camera is zoomed in for the close-up of his bench material. If she relied only on her viewfinder monitor she would be unaware when the demonstrator had shifted his gaze to the lens in order to make a teaching point with his face in vision. Compare this with 16 and 17.

The primitive cameras and mounts shown in 13 are in use with a programme monitor shared between the demonstrator and the

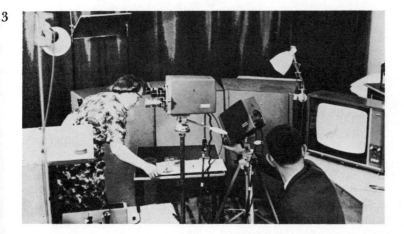

operator, who here must rely on the programme monitor for all adjustments he makes in camera or lens angle, once the programme has started. So it is important to check focus and measure distances accurately beforehand. The use of the turret in Camera I makes this easier, because each lens can be corrected for an appropriate shot, and the turret rotated, out of vision, during the programme in order to bring the appropriate lens to bear on subject matter (placed on the display platform to the left of the picture). Alignment of Camera II, which uses a supplementary lens for close views of pond life in the saucers on the demonstrator's table, must be done very carefully before the programme begins. The demonstrator moves in each dish to a marked position on the table as required. There may, however, be variations in the size, and brightness of the subject matter, which requires adjustments of Camera II's zoom lens during the programme; relying on a single monitor means that its preliminary adjustment must be carried out in the intervals whilst Camera I is transmitting, and is therefore done blind—without reference to a screen.

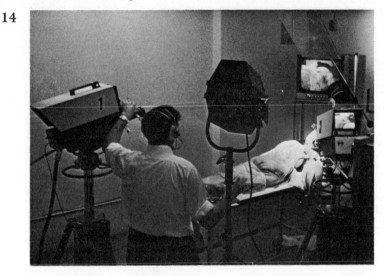

One operator is using two cameras, one pointing at an overhead mirror (see also pages 48–59), the other giving an oblique close-up view of the subject. The distances from lens to subject were checked before the programme began, and the focus set. There are no substantial changes in the brightness of the subject matter, and this means that the aperture can be set beforehand too. This leaves the operator free to adjust zoom on either camera as the programme requires, acting on guidance he receives from the programme director who operates the vision-mixer (see 19). Where a major change in the zoom must be made between shots, the operator makes use of his viewfinder monitor. Otherwise, zooming in vision, he can do just as well by relying on the programme monitor which the demonstrator is using in **58**. The demonstrator has moved out of the way for picture 14 to be taken, which shows how easily the operator can move from one camera to the other, and how he relies on the programme monitor. (A second programme monitor, behind the operator, is available to the demonstrator if he should wish to move round to the other side of the operating table.)

Operation of the vision-mixer is by touch, in order that the operator can keep watch on the subject matter. Her hand rests on the panel of the mixer, so that her fingertips caress the switch buttons, and cuts can be made instantly without delay in locating the button (15).

15

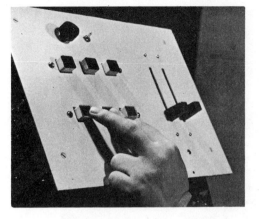

Compare 16 with 12. Here the operator has two cameras to cover the action. She has aligned Camera I for a close-up of the demonstrator's head and shoulders, and Camera II provides the close-up of the model he displays. Once again she momentarily ignores her viewfinder monitor in order to see what the demonstrator is up to. Her right hand is on her vision-mixer panel, so that the moment the demonstrator deflects his gaze from Camera I's lens to the model boat in his hands, she can cut to Camera II.

16

17 shows a similar situation, in which this time the operator is manipulating the faders (on the right of her vision-mixing panel),

in order to mix backwards and forwards between two sets of captions to show a series of map changes. Here the gaze of the operator is directed neither to the caption manipulator nor to her programme monitor. She must not begin her mix until the caption covered by the camera on her left is in position. Her viewfinder monitors single out the relevant captions, and it is less distracting for her to rely on the appropriate viewfinder monitor than to look direct at the caption manipulator whilst trying to remember which caption is "in vision."

The caption manipulator must not initiate a change of caption until a mix has been completed. The only way he can be sure of this is to watch the programme monitor. In between mixes he may look down at his caption racks, so as to grasp the tab at the side of the next caption to be removed.

External Control

The vision-mixing trolley (18, compare 17) can also provide for remote direction, either in a separate cubicle or, on location, in another part of the same room where the cameras are at work. The three screens on the shelf correspond to the viewfinder-monitors of the three cameras in use on the floor. The screen mounted on the vision-mixing trolley remains the programme monitor, showing whatever picture, or combination of pictures, is being relayed for the viewer. A hand-held microphone enables the

18

operator to direct her cameramen, via the headphones each will wear. Compare this with 20 in which the director/vision-mixer is using a swan-neck microphone fixed to the trolley, so that he can use both hands on the vision-mixing panel, left hand poised over the switches, right hand grasping the faders.

In 19 a director does his own vision-mixing, and also speaks a commentary. He is linked by the microphone suspended on a lazy-arm stand with the operator (who is shown in 11 and 14). The director wears headphones so that during rehearsal either the operator or demonstrator can tell him of problems as they are encountered. During the final transmission the director will set the

9

pace of the demonstration by the commentary he gives; the demonstrator can overhear this via the headphones he wears (see 58). The director will divide his attention between the monitors and the commentary notes on his clipboard. The smaller monitors represent the three cameras which he can command, one of them linked to a microscope. The larger monitor shows the programme the viewers will see. During preparation and rehearsal he will be concerned rather more with what individual cameras have to offer. During transmission, having agreed all the shots beforehand, he confines his attention to the programme monitor, and this leaves him opportunity to consult his commentary notes.

The director in 20 is using his microphone merely to communicate with his camera operators. Any live speech will come in from the studio. The director keeps his eyes on the monitors almost all the time, because the camera operators depend on him for guidance both in aligning their shots before he cuts to them and in making camera movements, or manipulating the zoom lens, "in vision."

20

A long and complex presentation is exacting for the director, if he is following a prearranged script rather than improvising a programme from the observation of unrehearsed activity. He relies here on a production assistant to keep him abreast of the script. Whilst his eyes are on the screens for most of the time, hers are mainly directed to the script, and she follows it with her finger so that at any moment the director can snatch a quick glance and instantly find his place. In addition she will make notes of timings in the margin of her script, and may also jot down any adjustments the director wants made after the rehearsal is over.

3
Hardware

The Patch Panel

The patch panel provides a cheap and versatile way of routeing vision and sound between one piece of equipment and another to suit differing requirements; to link the source of programme

material to the studio monitor that is most conveniently placed for performers or caption manipulators; to involve one or more video-tape-recorders or audio-tape-recorders, either to receive programme material, or to replay material as an insert in the programme; to distribute live or recorded material to different audience areas; to feed back comment; to record radio or television network transmissions.

Ease and efficiency in operation depend largely on the clarity with which sockets are marked, and distinguished. It may be useful to introduce a colour coding to distinguish audio, vhf and video distribution. The sockets should be placed far enough apart for each identifying label or number to show up distinctly.

Leads when not in use should hang clear, so as to avoid unnecessary clutter.

Sound

There are more problems and possibilities to be encountered in the effective use of sound than in that of vision. Those who doubt this should investigate the varieties of microphone already on the market, and try to pick a good all-rounder. Or they might blindfold themselves and listen hard to all the sounds that go on within earshot. The problem with microphones is that they are both more and less discriminating than the human ear. A great deal of what we think we hear through our ears is in fact being interpreted to us through our eyes. At a party one may be listening "with half an ear" to one's companion, whilst questing for someone more interesting whose voice can be identified even though it comes from the other side of the room. One can shift one's attention and phase out the near voice in favour of the distant voice, without stirring an inch. The microphone, unfortunately, picks up the lot, but with a more limited frequency range, so that the convivial sound of the party emerges through the loudspeaker as mere noise.

The first job is to pick an acoustic situation in which there is as little competition as possible with the sound you want to obtain. Competition can come from the distorted reflections of that self-same sound that are bounced off hard, smooth surfaces; or it may be conveyed by the structure of the building itself—along corridor, or ceiling girders, or by hot-water piping through the walls—from somewhere far away. Or the sound may suddenly arise near at hand when people pour out of an adjoining room into the corridor or when rush-hour traffic congests the road outside.

This is the quietest corner of a large room. The window overlooks a patch of garden, not a street. The microphone is inclined
away from the echoey interior. Curtains and table-cloth help to
damp down sound, and an additional sound-baffle—a blanket
thrown over an easel—has been introduced. Besides absorbing
sound this is also useful in reflecting light (see pages 96 and 97).

Choose microphones to suit not only the surroundings but also
the variety and span of sound sources to be included. A "gun"
microphone helps to isolate an individual speaker in the open air,
or in a large hall, where there is a teach-in for instance. A much
wider angle of acceptance is required for a general discussion, or
for the reproduction of appropriate background sound. Often the
best results are obtained when a microphone with a fairly narrow
angle of acceptance is held by one of the participants—interviewer.
chairman, commentator—and pointed in turn at each as required.
This technique is particularly useful in a free-for-all interview or

24

discussion where voice levels vary, and the participants may lean forward to make a telling point. The chairman or interviewer learns to vary the distance between microphone and speaker to allow for different voice strengths, and to incline it towards each speaker as the talk goes to and fro.

25

26

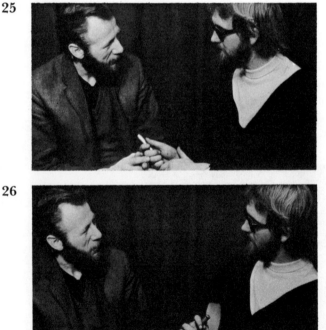

The most robust and versatile microphone is a *moving-coil*: held in the hand, used on a table stand or floor stand, suspended from a boom, or clipped to the side of a wall cupboard. At a pinch

27
28

it can be used in the halter position—though this lightweight *dynamic* microphone is to be preferred because it is more comfortable and less conspicuous.

When the halter microphone is connected by a lead to the sound system it tethers the presenter unless he takes it off to move further afield. It should then be hooked up safely within easy reach, on the presenter's bench.

29
30

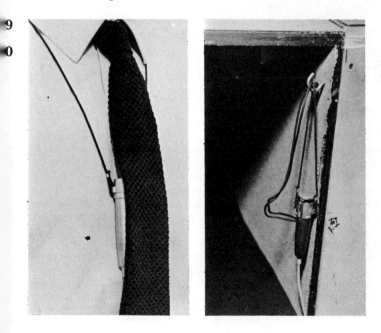

Alternatively, use the same type of microphone as a *radio microphone*. A compact transmitter fits into the presenter's pocket, or is slung from the belt, and relays sound to a tuning unit attached to the sound-control point (usually beside the vision-mixer).

31

The diagram shows some basic considerations governing the choice of any microphone: impedance, angle of acceptance, frequency response and sensitivity.

32 MICROPHONE TYPES (page 31)

Crystal: Omnidirectional; low cost; 30—7000 Hz. Robust; poor treble response; use for close voice work.

Moving-coil: Omnidirectional; medium cost; 50—13,000 Hz. Robust; use for interview, including out of doors, and for music.

Ribbon: Two "live" sides—figure-8 response; medium cost; 50—15,000 Hz. Fragile; must be kept still; use for indoor interviews, music.

Cardioid pattern: "Heart-shaped" pattern achieved by a composite of ribbon and moving-coil microphones. Medium to high cost; 50—15,000 Hz. Use for interviews or music; reduces background noise; can be used as boom microphone.

Gun: Narrow acceptance angle; expensive; 50—12,000 Hz. Bulky but manœuvrable to stay with moving sound source; used with windshield out of doors; also good with indoor audience participation.

Lanyard (Moving-coil): Usually worn round neck; medium cost; 6000 Hz peak as microphone is off voice axis; uses chest as a resonator; poor bass response helps to compensate for clothing rustles. Small, light and robust.

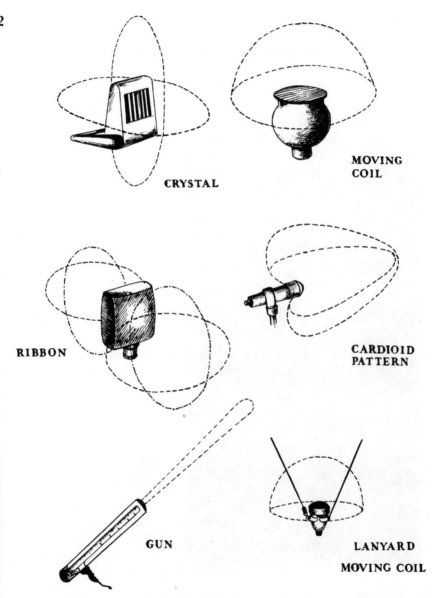

CRYSTAL

MOVING COIL

RIBBON

CARDIOID PATTERN

GUN

LANYARD MOVING COIL

This *ribbon* microphone accepts sound from back and front, though manufacturers sometimes stuff one side with sound-absorbent material in order to reduce the incidence of unwanted sound. The sensitivity of the ribbon suspended inside the protective casing demands use in a fixed position, either on a stand or slung from a boom. If it is moved whilst sound is being relayed there may be unwanted noise from the passage of air as it moves.

33

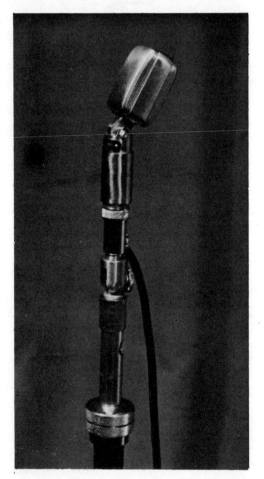

. The angle of acceptance of the ribbon microphone is fixed, but in the more complex *cardioid* microphone the angle can be varied by a switch which brings in a second component.

Video-recorders

The first video-tape-recorder to come within the budget usually available to CCTV users was the Philips one-inch ETL 2700, running at first at 7½ in. per second, subsequently upgraded to 10 in.

34

per second, and surviving to do reliable service for many years afterwards. Whilst the tape is moving comparatively slowly from one spool to the other the head unit revolves at great speed, rising up and down as it does so, within a loop formed by the tape, called the *wrap*.

Other machines, such as the Ampex range, used the same *helical scan* principle. The price paid for this economical system of spooling the tape is usually shown in *wear*, on the head and on the bearings which guide the tape on its path. With the *alpha* wrap the tape is carried upwards round the head and then downwards and on to the take-up reel. The torsion this involves inflicts wear on the tape—in some machines this deposits oxide which rapidly clogs the head.

Frequent cleaning of the bearings is therefore important. The spools are removed.

5

And so is the collar which protects the clean unit.

36

A metal-cleaning solution is applied with a brush reserved only for this use.

37

Each roller on which the tape bears is wiped carefully (38).

38

The recording head needs particular care (39, 40).

39

40

The rubber capstan which moves the tape on its way, when the clutch mechanism is engaged, should be wiped with a clean dry cloth (no solution).

41

More recently one-inch recorders such as this IVC have been designed with the tape wrapping round the head on an *omega* wrap, which reduces head wear, and clogging. Cleaning procedure is simplified.

42

Protect the video-tape-recorder from dust. The best machines have their own Perspex covers, which should be kept closed most of the time unless it is impossible to reach controls otherwise than by opening.

Once the video-tape-recorder has been cleaned, the tape laced, and the footage counter set to zero, make test recordings on the output from each camera in use. These may be no more than the test-card captions shown on page 11. It is sometimes also useful to check recording levels on particularly contrasty subject matter, where the balance of dark and light may be a question of aesthetic rather than engineering judgement.

43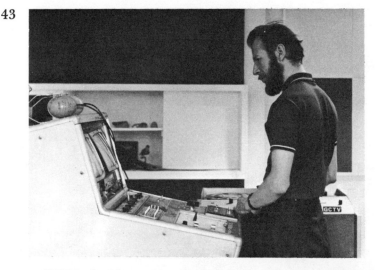

Half-inch video-tape-recorders have quickly proved their worth, and the designs now available combine ease of operation with minimal maintenance. Colour recording is practicable with the more expensive models.

44

The greatest advantage the half-inch machine confers is in its portability. Some of the possibilities this opens up are described on pages 110–17. Lightweight machines can be operated by switches linked to a trigger grip on the camera. This means that, once loaded, the video-tape-recorder can remain closed throughout the operation.

45

Editing of video-tapes is best done by copying a new sequence on to a tape from the point at which the editing is intended. For example, if a programme has been carefully rehearsed and a mistake is made during the recording, you may decide to stop the recording, spool back to a convenient point on the tape before the mistake occurred, and then set the video-tape-recorder in readiness, but with the clutch disengaged.

The performers and the camera operator(s) go back to a point even earlier in the programme sequence, and resume. By the time they reach the point at which the recording is resumed, their pace will be back to normal.

Much depends, therefore, on choice of a convenient point on the tape at which to re-record. If possible, choose a static subject, such as a caption or an object shown in close-up, where there is a pause in accompanying speech. Set the tape at the point on the recording just *before* the cut is made; run the machine just *after* cutting to the caption.

Another reason for editing is to compile several extracts from different spools, or from different sections of the same tape, by copying each section, in the required order, on to a new tape, on a second machine. This introduces a further complication—possible differences in the video patterns that have been recorded on the original tapes. On some machines it may take a few seconds for the patterns to settle down and conform, giving a fireworks effect on the screen. Eliminate this by completing the copying of each section with a fade to black. Set up the next section a couple of seconds in advance of the first scene required, so that the re-recording can fade up from black.

The more sophisticated machines, when accurately maintained, will provide smooth cuts, without any time-lag while the sync. pattern settles down.

As a last resort, edit video-tape by cutting, and joining the chosen sections with splicing tape. This becomes easier, and more effective, with a narrow-gauge tape. The angle of the cut should be diagonal, in order to allow for the difference between the position of the video signals and the audio signals.

Editing by cutting continues to be undesirable with video-tapes, because of the high speeds at which the record/replay heads must scan them, and the resultant strain on the tape, and risk of abrasion to the head by the join.

Audio-tapes are another matter. They deserve a section to themselves (pages 101—6) because the tapes are relatively so cheap to buy, so easy to cut and rejoin, and because the finished product can always be copied on to the track of a video-tape, either simultaneously with the recording of the visual or subsequently.

Pan/Tilt Heads

At its crudest and cheapest, the pan and tilt head may consist simply of a hinge mounted on a bearing. The hinge gives the tilt movement, and is here locked by rotating the pan/tilt handle. Movement in the horizontal plane is achieved by rotating the base of the pan/tilt head (arrowed) on the top of the tripod shaft. This movement can be locked by a second screw, but there is then a danger that the operator will inadvertently attempt a panning movement without unlocking. The result may then be that the pan/tilt mount begins to unscrew from the shaft of the tripod. To avoid this danger, the locking device has been deliberately removed in the picture shown.

46

A home-made counterbalance frame has been added, in order to keep the camera in balance at whatever angle it is tilted. The lead weights on the base of the frame can be varied in number in order to counterbalance the combined weight of the viewfinder monitor and the lens.

At the other extreme, the pan/tilt mechanism is balanced hydraulically, with adjustments to height as well as angle when required.

A useful compromise is the cradle system (see 4 and 11), although it may not quite eliminate unwanted movement in the vertical plane when the camera is unlocked and unattended. Movement up and down is eased by a spring-loading system within the central column of the tripod or pedestal.

The spring loading must be adjusted to suit the weight of the camera plus lens plus viewfinder monitor, if any of these items are changed from one production to the next.

The most recent development is this "camera station," which combines very smooth and adaptable pan/tilt movements, finger-tip control, and great stability on a specially sprung wheelbase. As the metal used is magnesium alloy, and each section is quickly demountable, it is quite easy to transport and manœuvre.

47

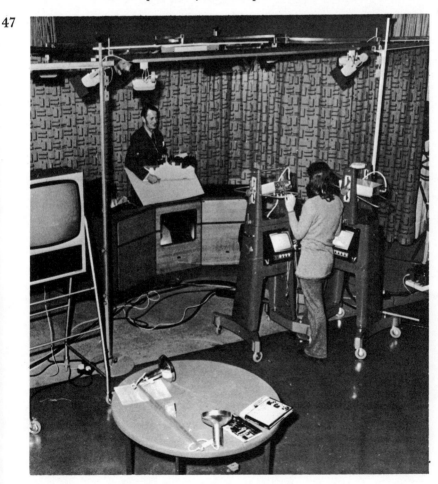

Remote Control

Remote-control housings are available to suit most cameras. One section holds the camera and encloses motors which control the tilt movement at *A* and the pan movement about the base at *B*. A separate unit, with its own motors, encloses the lens, and provides independent control for focus, zoom and diaphragm.

48

The complete unit may be placed on the floor, to get an upward view of a ballet movement; or on the top of a cupboard, to look down on a workshop or classroom. It is important to allow enough free cable to permit full movements in both the horizontal and the vertical planes. The cable here has been taped to the back of the cupboard at *C*, so that it exerts no drag on movement.

Studio uses of a remote-control camera, placed on the top shelf of a trux (combined storage unit and acoustic screen) looking inwards on to a set in the foreground, or peeping through the gap between two trux. Note in each the careful arrangement of the cable.

49
50

A variation is this remote-control camera mounted on its own pedestal, in a lecture room or workshop, but manipulated from far away.

51

Special Effects

It is sometimes assumed that these consist simply of what an *effects generator* can provide. This is a device that, coupled with the vision-mixer, can produce wipes, inlays and vignettes electronically. In practice it can be surpassed, or bypassed, by the imaginative use of less costly visual aids—pictures, slides, film clips —always provided that there is the right kind of simple but effective ancillary equipment.

In 52 is shown a primitive arrangement of *coupled slide-projectors*, one above the other in the foreground, plus a *cine-projector* whose beam is reflected on to the screen by the

52

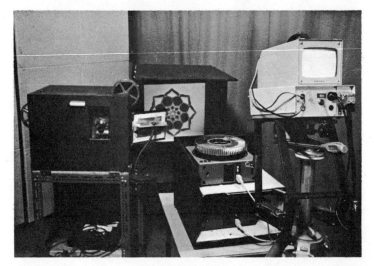

adjustable mirror. In this way three separate images can be projected on the same screen, if required, at the same time. In practice it is rare to use more than two images at once, for example a word caption on a slide, labelling an important element in part of the sequence projected on film.

The two slide-projectors alone are good value as a device for mixing from one slide picture to another, or superimposing one over the other. The slide changes are regulated either by a lever whose operation slowly opens one aperture as it closes the other, or by a rheostat which dims one projector as it brightens the other. Whatever the combination of images, the camera is there to

relay the result, and the only adjustment it may require from time
to time is on the diaphragm, to allow for variations in the light-
intensity of the images projected.

The screen is enclosed in a wooden box to keep out unwanted
shafts of light, and may be viewed, as shown, from the front; or,
by the substitution of a back-projection panel, from the other
side.

53 shows the way in which the same basic principles have been
used with much greater professional finesse. In this *multi-purpose
display unit* there is room for two projectors (either two slide-
projectors, or one slide-projector and one cine-projector) on either
side of the base of the column, up which their beams are deflected
by periscope mirrors. The image is shown on a horizontal back-
projection screen which in turn is viewed by the camera through
an inclined mirror. The great advantage here is that the same
camera arrangement suffices for three other types of shot. It can
be used for animation against a projected background; the screen
can be covered with a small mat, on which an object or a caption
can be placed and lit from above. The camera views the subject *in*

(a),

(b)

53(c)

53(d)

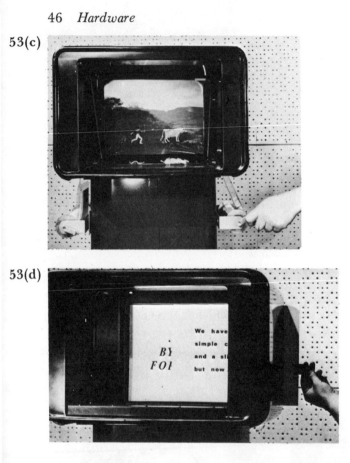

plan through the inclined mirror. By zooming out and panning down the object can be shown simultaneously *in plan and in elevation,* as on page 85. Having panned down the camera can be zoomed in to show the object in *elevation* only.

The inclined mirror can be dropped flat, so that the camera can view a sequence of captions stacked at the back of the demonstration box. A spring device ensures that each caption is kept firmly in place.

One advantage of the demonstration box is that it can make effective use of small captions. When the camera is fitted with a supplementary lens, it can command a full-size 12 × 9 in. caption, or it can be zoomed in to show the detail within a picture post-card. A stack of plain postcards will serve for captions without the need for photographic enlargement.

The bench frame containing the *underneath mirror* offers another range of subject with no need to change camera position. The inclined mirror shows the camera whatever is displayed on, or above, the bench top. The glass sheet which forms the bench top

4

can be lit from above or below or both at the same time. Besides model work, a caption can be viewed face down, or cut-outs shown in silhouette, or a sheet of translucent paper used as a writing surface.

Some of the many possibilities for elaboration are described in the next section.

4
Angles and Layouts

It's All Done by Mirrors

The effectiveness of the underneath mirror frame depends on exploiting the scanning system in the camera, described on page 5. Normally the dot traverses the screen from left to right and from top to bottom, to scan the picture. We found, early in the first Hertfordshire television experiments, that it cost little to add switching devices to the camera which could reverse the direction of the scanning. *Reverse line* makes the dot travel from right to left, and the effect is to turn the picture round, so that a mirror image can be corrected. *Reverse frame* causes the scanning to go from bottom to top, and thus turns the picture upside down. It is used with the underneath mirror frame to compensate for the fact that the camera is on the opposite side to that occupied by the writer.

Given these two devices, which many manufacturers nowadays include in their cameras, or which can be added at the request of the purchaser, the limitations of space, equipment, and manpower under which closed-circuit television operators often labour can be largely overcome.

In the simplest form, mirrors provide new camera angles without the need for the camera and operator to take up new positions.

The worm's-eye view afforded by the underneath mirror frame is a useful way of studying objects or insects resting on the glass, and lit from below. Manipulations such as the assembly of small components with a pair of tweezers can be viewed from underneath, where the craftsman's hands and tools do not obscure what he is doing.

For many craft demonstrations, however, the most significant viewpoint is, as it were, through the eyes of the expert, so that the viewer learns to manipulate tools and materials himself by matching the view he has of his own hands with the view he remembers of the way things should look when an expert handles them. The *overhead mirror* can show us substantially what the demonstrator sees as he displays his materials, or pursues his craft, or conducts his experiments. In order to avoid the lino-cutter's head coming into shot we must accept a slightly more perpendicular view via the mirror than she has when she views her work direct, and it is

55

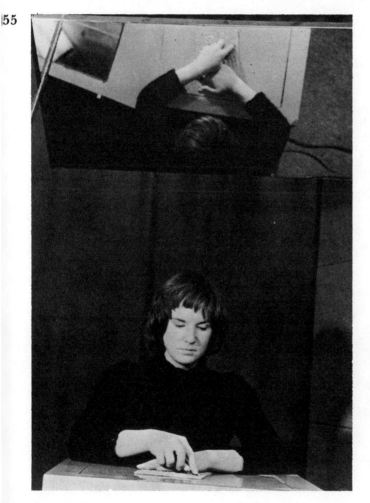

usually good policy to suspend an overhead mirror slightly forward
of the demonstrator, as with this surgeon dissecting a brain specimen. But his subject matter shows to advantage, and he can use
his own monitor, on the bench beside him, to check his movements on the image corrected from the mirror so that his viewers
see what he wants them to see.

56

57

Sometimes an oblique view is more revealing, as in this anatomy demonstration where the mirror overlooks the specimen on the table from one side of the surgeon/demonstrator.

58

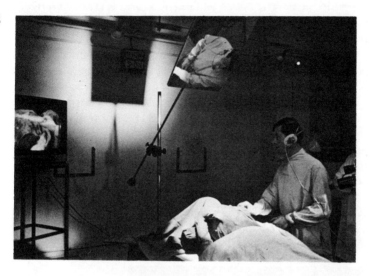

It is for the operator to decide the best mirror position, having first looked at the job to be demonstrated from the standpoint of the viewer, and afterwards checking the most promising mirror angle from the standpoint of the demonstrator to make sure that he has not created fresh problems—unwanted shadow or a mirror set too low for a tall demonstrator who must also be looked at by the camera direct when he is talking to camera.

Easy manipulation of the overhead mirror is achieved by using a mirror that is light to handle and can be suspended without the need for a heavy framework to hold it in position. The answer is a silvered fabric stretched over a lightweight aluminium frame. The material was originally manufactured to provide mirrors for toilets in aircraft. It was first used for television in the Hertfordshire project, and has spread to many other television installations since. Mirrors can be made up by the manufacturer to suit the size the user specifies, and the type of fixture he finds most convenient. Because there is no layer of glass above the silvering there are no refraction problems when the mirror image is viewed from an awkward angle.

Care of the mirror fabric is important, but not unduly exacting.

Provide covers made of hardboard with a soft foam plastic lining to protect each mirror in transit. Clean by swabbing with the cleansing liquid recommended by the manufacturer. Protect the most vulnerable parts—the edges where the fabric is stretched over its framework—by adding a binding strip of adhesive tape, giving a ½ in. margin all round the mirror. Occasionally, if the fabric has not been accurately stretched over its frame, a slight crinkling effect may appear, or there may be an almost imperceptible sag in the surface. This can usually be corrected by warming the surface —for instance by holding it close to a lamp. Warmth causes the fabric to contract, instead of expanding, and the unevennesses are stretched smooth again.

The same mirror, equipped with a versatile fixture on the back, may be used in many different situations: suspended on a boom, supported by a floor stand, levered from a lug in the ceiling, clipped to a wall bracket or one of the multi-purpose trux illustrated on page 43. Besides the demonstration of manual skills, or of objects and equipment shown in plan, a mirror may be used for reverse angle shots, of this castle model, for example; or side views

59

60

where the mirror also helps by reflecting "filler" light on to the presenter's profile.

61

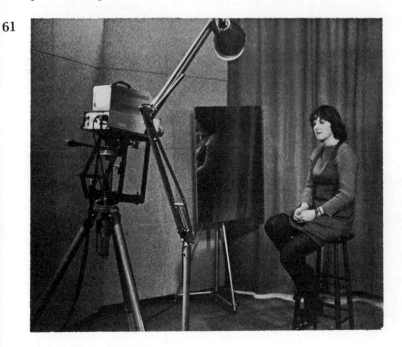

A simple display stand, home-made, provides a working surface on which to show objects in plan, or captions face up, or a hand at work.

62
63

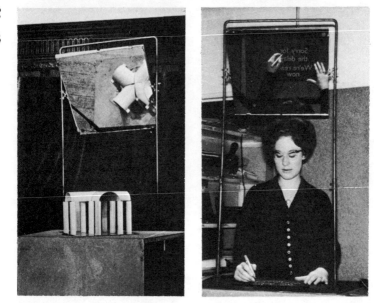

A variant on the display stand is the frame below, which can be attached to a draughtsman's easel (shown on page 89 in use as a simple display surface). The angle at which the object is shown to the camera can be adjusted either by varying the angle of the mirror on its bar or by tilting the easel (with mirror support attached) on its stand. This allows for adjustments to make the best of the available lighting.

64

This boom slides up and down a floor stand, and angles can be adjusted both on the shaft and on the universal joint attached to the back of the mirror.

65

A telescopic boom makes it possible to swing the mirror from one working area to another as required.

66

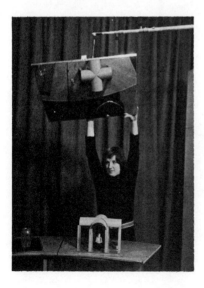

Adjustment of the mirror angle is best done by aiming the camera at the mirror, zooming out (after checking that there are no bright lights in shot) and then switching the camera through to a monitor that can be seen from the area the mirror occupies. Go to this area and make fine adjustments on the mirror by watching the effect on the monitor. In this way you can make sure that the mirror reveals the subject matter from the appropriate angle, that it does not pick up unwanted reflections of bright light, and that the demonstrator's head does not get in the way when he gets to work. Finally, go back to the camera, pan downwards (and de-reverse) and check that the positions of the mirror and the lighting do not combine to cast an unwanted shadow over the background where it is in view for a normal shot.

The plan diagram shows how a mirror may be slung from parallel cords stretched between hooks, wall-bars, or girders that are part of the existing fabric in a room that is already too congested (with factory machinery or nursery school toddlers or vaulting gymnasts) to permit of the use of a floor stand or a telescopic boom. Horizontal cords slung from irregularly placed wall or ceiling fixtures support a frame within which the mirror can be angled to suit the camera.

67

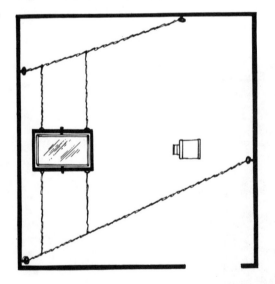

A very small mirror is extremely useful in close-up work using a supplementary lens (see page 90). An overhead view sometimes

allows the subject matter to be more easily lit, and where a specimen is immersed in liquid—the pond life in the saucer—nothing
needs to be upset in order to gain a commanding view. This technique is particularly effective in the multi-purpose unit shown on
page 85, where the camera can be panned and zoomed to give an
overhead and/or an elevation view of a small object which can also
be illuminated from underneath.

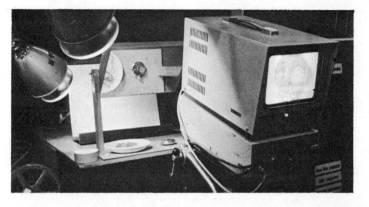

Use of the underneath mirror frame to display objects from a
worm's-eye view, such as the model on page 47 to demonstrate the
principles of barrel-vaulting, seldom requires special adjustment,
since the underneath mirror is angled to reflect the underside of the
glass bench top towards a camera at normal operating height,
placed at a distance which also conveniently commands a front
view of the subject matter when required.

It may sometimes be necessary to move the frame containing
the mirror in order to avoid getting a distracting view of apparatus

suspended from the ceiling. Compare this with the clear background shown in the mirror on page 47. We can, of course, ignore
such background problems in most types of over-drawing and
animation (see pages 59–60), where a sheet of diffusing tissue or a
box hood intervenes between the mirror and the ceiling high
above the frame.

70

Both the mirror and the glass sheet inset in the frame need routine
cleaning. Note that an ordinary silvered glass mirror can be used instead
of a fabric mirror in the underneath mirror frame, where the
danger of smashing and the problem of weight in suspension are
much less than with overhead mirrors.

Remember, whatever mirror you use, that when focusing the
lens the distance that matters is not X-Z but X-Y-Z. This situation
can be exploited by placing other subject matter which the same
camera will cover at other times during the programme at the
same distance, X-W, in a straight line from the camera lens. In this
way one can exploit the hub and arc principle described on page

71

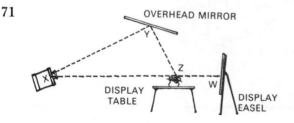

OVERHEAD MIRROR

DISPLAY
TABLE

DISPLAY
EASEL

12, and keep objects in other display areas—the wall behind the demonstration bench, or an easel to one side—within the camera's reach without the need to readjust focus.

Use of a supplementary lens (see pages 88—91) for very close work will require a change either in the angle of the mirror or in the relative heights of mirror and camera. In practice it usually suits best to raise the mirror height by putting the frame on a box rather than by lowering the camera height to a level which the operator finds awkward.

Animations

Some of the most sophisticated effects in television can be achieved by the simplest operational means. It takes little trouble to line the camera up on the top mirror or the underneath mirror frame, and leave it. The simplest and most effective animations require merely that the edges of the working surface viewed by the camera are not visible to the viewer. In this way a sheet of card (face upwards beneath the top mirror or face downwards over the underneath mirror) can be moved about freely, and provided there is no distinctive pattern to its surface it can have a small symbol, drawing or even a small object drawn on or stuck to it which will appear to move about of its own accord because the camera does not reveal that the whole of the background is moving with it.

Use of the underneath mirror frame makes it possible to interpose a piece of foreground "scenery" in the form of a cardboard

cut-out or, as shown in 72, some natural materials to suggest vegetation on the sea bed. A single fish can be made to move about by moving a dark card with the fish on it, over the surface of the glass frame. If more than one object is required to move, separate cut-out symbols can be manipulated by black-gloved hands. The photograph shows them without the box-shaped hood which is placed over the bench to cut down the lighting from above so that the black-gloved hands disappear and only the cut-outs they manipulate show up.

73

74

A more traditional technique depends on sliders sandwiched between two sheets of card, one of which has holes or slits through which the operation of the sliders can be seen. This young lady's eyes and mouth open and close by pulling one or other of the tabs which project below the area the camera views.

75

To get the right conditions for successful animation, the operator must make sure that the manipulator has a good view of a monitor from where he works. The card has been mounted so that whoever operates it may sit below and looking upwards be able to see both the card and the monitor.

Captions

Some of the best programmes are put together in a hurry, whilst enthusiasm is at its peak. An outside contributor has been coaxed into lending a hand or taking the principal part. To make the most of his ideas requires very careful marshalling beforehand of those resources which can be quickly exploited to illustrate the points he wants to make. Two-dimensional pictures are often the answer —provided a good selection is available from which to pick and choose. A store of possible illustrations should be accumulated, and filed, from the moment a television unit first sets up shop.

This rack contains cut-outs from magazines, photographic enlargements used in previous assignments, sections cut from travel posters and photographs clipped from newspapers. The boxes on the top row contain unmounted cut-outs, whilst the divisions below house captions that have already been mounted. There is a decimal filing system, with a key to it attached to the side of the storage rack.

76

Wherever the size allows, the pictures are mounted on standard cards measuring 12 × 9 in. It pays to concentrate on this standard size, which can be screwed to the wall.

77

A selection of other types of caption rack should be available, to contain outsize pictures and charts. A "lazy susan" is also useful

78

to display an abstract, or an object, and rotate it in the horizontal plane for a special effect. The domestic circular cheese-board is a useful substitute.

The *aspect ratio* of the television screen imposes limitations which must be borne in mind when preparing a caption card. The camera only shows pictures in the proportion 4 (length) by 3 (height). If the subject does not correspond to these dimensions there are two ways of coping with the problem. One can look into the picture, allowing the camera to pan over the subject matter, showing as it were a series of "pictures within a picture"—close-ups and medium shots, always in the correct aspect ratio, within a larger picture which may not necessarily have the correct aspect ratio. The alternative is to choose a sheet of caption card of the correct aspect ratio and mount the picture on it, so that there is a neutral background (grey or black, not white) on either side. Whichever the method a *"ratio ruler"* is a useful aid. Cut out L-shaped strips of cardboard consisting of 12 divisions along each axis, with the vertical divisions three-quarters of the horizontal divisions. Place two L-shapes together to make a rectangle. The size of the rectangle can be varied by allowing the L-shapes to close in on each other so that there are fewer divisions on each axis. So long as there are the same number of divisions on each axis the ratio of width to height remains the same—four : three. A pair of ratio rulers laid on the surface of a poster, or on any other outsize caption, can be used to frame a chosen detail, and to determine what else will necessarily be included in the television picture in order to span the chosen subject.

Within this frame allow a generous margin (up to 1½ in. on a 12 × 9 in. caption). Otherwise you risk losing important information because someone's screen does not quite correspond to the area on your own viewfinder monitor. A picture caption will, of course, "bleed" to the edges of the card wherever possible, but the significant content should be within that imaginary wide margin. It is better to spread verbal information over a series of captions or charts rather than try to pack it all into one caption.

Bold effects are usually best. Like a good designer of road signs or posters you are aiming at impact within a short measure of time. Choose thick, spacious typography. Experiment with the layout and with the tones in a rough version in order to try out the effect on a camera. Diagrams, charts, maps and word captions go well in one of the following combinations:

White and black on medium/dark grey card.
White and grey on black card.
Black on grey card.

The easiest way to improvise and adjust is to lay caption
material on a horizontal surface (where gravity does the sticking)
and view it through an overhead mirror. This is particularly
important where an illustration, or a diagram or a word is to be
superimposed on another subject shot by a second camera. Line up
the second camera, superimpose Camera I and move scraps of card
or paper bearing the picture, diagram or word until they occupy,
in superimposition, the most convenient areas in Camera II's shot,
prominent but not obscuring the detail to which they refer. This
procedure will save you from the fate of the bad foreign-language
film-titler who always seems to get white lettering superimposed
on a light patch in the picture. Once a satisfactory arrangement of
the caption material is achieved, the final version of the caption
can be made; or as a time-saver you can compose your caption
material on cards stuck to strips of magnetic tile, placing them on
a matt black background backed by a metal sheet. This allows for
last-minute adjustments without the need to remake the caption.

Diagram captions where a very precise superimposition is required can sometimes be assembled more conveniently on the
glass sheet inset in *the underneath mirror frame*. For example,
suppose you want to show, simply, the path followed by a piece
of raw material through the cogs and sprockets in a complex
machine. Having lined up the second camera on a profile view
of the machinery, cardboard is cut out in shapes to correspond
with the aspect of the cogs as they lie within the surrounding
apparatus. Camera I zooms in or out in order to get the proportion
right. The cogs are pushed about over the surface of the glass
until they are positioned correctly in relation to Camera II's shot.
If the diagram is not required to move, all that is necessary is to
place a black background card face down over them and to tape it
or weight it firmly. If a moving symbol is required, for example
something to represent the raw material as it passes through the
machinery, this can be stuck on to the background card, which is
then held free and manipulated in order to provide an animation
of the kind described on page 59.

Over-drawing is a simple use of a similar technique. The demonstrator wants a rapid means of drawing attention to some feature
of an object, caption, slide or film sequence which is being shown

to viewers via the other camera. He decides whether to write in white letters on a dark background or in dark letters on a light background, in order to achieve the necessary contrast with the predominant tones of Camera II's subject matter. If he chooses to make his mark in white, he uses chalk on a sheet of black matt paper underneath the overhead mirror. He may be happy for his hand to be in view as he chalks, or he may prefer to conceal it by wearing a black glove, so that only the tip of the chalk shows up in superimposition.

If the most contrasting effect requires black marking, then the underneath mirror frame is used, with the glass covered by a sheet of translucent paper (tissue, vegetable parchment, or Kodatrace). Camera I views the underside of the glass top, via the underneath mirror, and corrects the image of the marks made by a felt pen (penetrating the paper and appearing on its underside) by use of both the reverse line and reverse frame switches. The demonstrator writes or draws in the normal fashion on the sheet in front of him.

Whichever writing surface is used, the technique required to match the word or diagram (on Camera I) with the reality (on Camera II) is the same. The demonstrator watches the studio monitor on which the superimposition of the two cameras is shown. He can see his chalk tip as he moves it about so he can identify it with whatever he wishes to arrow, label, outline, hatch in or otherwise indicate. Similarly he can see his hand in silhouette as it hovers over the sheet of tissue on the underneath mirror frame, and so can judge where to place the tip of his felt pen. Here he shares a monitor with the camera operator/vision-mixer.

79

In **80** and **81** a monitor is kept exclusively for the demonstrator, who here uses an overhead camera to display slides shown on his microscope. Note his eye-line to the monitor.

80

81

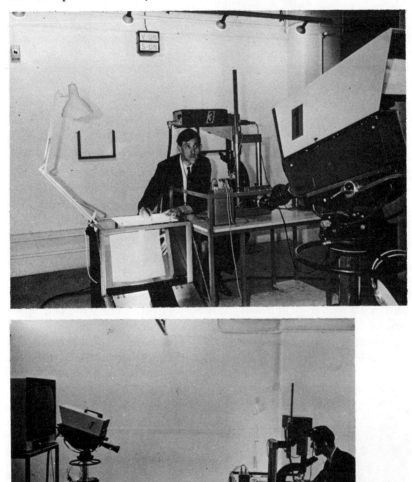

He sometimes uses his underneath mirror frame to display the kind of gelatine overlay originally associated with the overhead projector. But here he can superimpose his overlay, at will, on the specimen shown in his microscope (82).

82

Some demonstrators may prefer to prepare their drawings beforehand, by sketching the main outlines lightly on the paper with a 4B pencil, which will show up just sufficiently for the demonstrator. The pencilled lines will not penetrate the paper to be revealed by the camera; once the programme begins the demonstrator looks down at his drawings and inks them in. Meanwhile the

83

operator keeps an eye on the studio monitor to make sure that the drawing remains in register with the object to which it refers. This is a less satisfactory arrangement, in the long run, because all the operator can do to correct an error is to adjust either the angle of the camera or the zoom of the lens. A demonstrator who is unused to the process may insist on working, as he normally does, with his eye-line direct to his drawing material, but after he has played himself in it is a good idea to coax him into trying himself out, in a spare moment, using the monitor as his guide. Most newcomers expect this to be much more difficult than it turns out to be. Acquiring the knack can be compared to getting used to a driving mirror for reversing a car; in practice one usually grasps the essentials in a minute or two.

A series of similar captions, mixing from one to another and each introducing a slight modification on its predecessor, is a useful way to suggest change—for example, the expansion of a city, the erosion of a coast line, the ageing of a face. Mark out with tracing paper or a cardboard template the outlines which do *not* change and this will provide a means of checking the alignment throughout. See page 22 for the teamwork between operator and caption manipulator.

Telecine Sequences

Film is a useful ingredient in television, and it deserves attention in accommodating it to the medium. A sequence should begin and end with shots that go on for long enough to allow for a slight delay in switching over from studio to projection and back. It often helps to key in the filmed material by interposing a visual, such as a caption, or a close-up of some inanimate detail in the studio, which also appears just afterwards or just beforehand in the film sequence.

The timing of the film sequence should be noted on the storyboard the vision-mixer is using, together with a sketch to remind him of the last shot or two in the film sequence, so that he recognizes when the sequence is about to run out and is ready to switch back to the studio.

5
Presenter's Viewpoint

Seeing What You're Doing

The best way to get a good picture is to provide a monitor screen
on which the demonstrator can see for himself what he is display-
ing, and how to present it to advantage. Then you can leave the
picture composition to him!

   This anatomy demonstration desk relies on a camera held permanently above the working surface on which specimens are
manipulated. No operator is required during the demonstration,

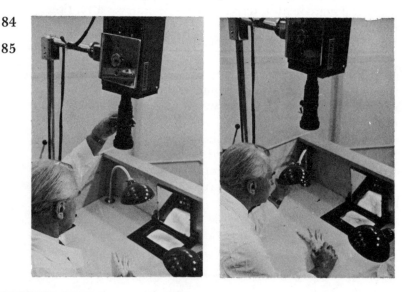

since both zoom control and accent lighting are within easy reach of the demonstrator (here the Professor of Anatomy). Even here a mirror is useful in order to show him his monitor screen, which is inset for compactness in the bench. The operator's attention is still required before the next demonstration in adjusting camera height on the ratchet control, and trying out alternative background materials to show a specimen in satisfactory contrast.

86

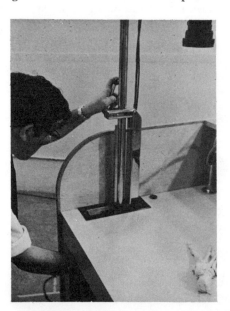

87

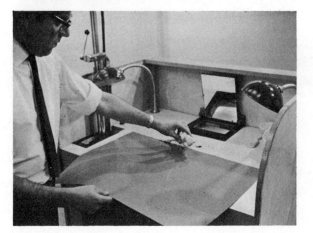

The surgeon here can judge for himself the effectiveness of pointer or forefinger as he indicates detail on the brain specimen. A pointer with one half dark and the other light in tone gives him a choice to suit different tones of tissue. The specimen is held in position with the wooden blocks, which can be moved around to give fresh support once dissection commences. The effectiveness of this overhead mirror shot for the viewer can be gauged throughout by reference to the portable monitor on the bench top. (See also 56 and 57 on page 50.)

88

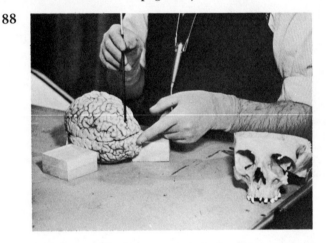

It is equally important to judge from a monitor whether the indicating finger is casting a distracting shadow, and whether there is irrelevant clutter (which might sometimes include the demonstrator's waistcoat) in the background. Here part of the display

89

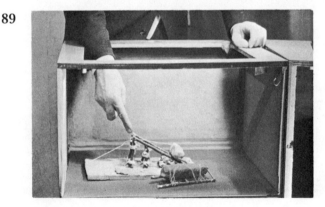

bench is turned round to provide an enclosed area within which the
model can be placed and lit, leaving the demonstrator free to move
about unseen. A section of the bench top is removed so that he can
point things out, looking throughout not downwards at the model
but forwards to the monitor on which he can see his forefinger as
the viewers will see it.

Over-drawing, already mentioned on page 58, is a more sophisticated way of indicating and/or delineating an object or a picture.
The device of superimposition may also be used more simply to
superimpose an index finger or a pointer (shown against a dark
background on Camera I) on a subject covered by Camera II. This
might be a microscopy specimen (see page 67), a frame from a
film strip, a cine sequence, or some three-dimensional object which
is too small, too remote from the demonstrator or too awkwardly
placed, for him to get at direct.

The position of the demonstrator's monitor needs careful consideration. His eye-line should need as little adjustment as possible
to go on a three-point path: to the camera lens, as he speaks direct

90

91

to the viewer in close-up; on to the subject matter he wishes to indicate, and then, as soon as he has located it and picked up the pointer or craft tools, onwards to the monitor to judge whether his demonstration makes satisfactory viewing. It is important, however, to save the demonstrator from the fate of Narcissus. If there is a large monitor in a dominant position beside the camera, it is all too easy for the demonstrator's eyes to shift to his own face on the screen when he should be talking to the lens direct. The answer is to keep his own monitor small and below the camera. If he drops his eyes towards the screen this appears to the viewer (ignorant of the monitor's existence) no more than becoming modesty; whereas a lateral shift of the demonstrator's eyes looks just shifty.

A larger screen can be tolerated if it has to be shared with the camera operator and is therefore placed at a convenient point where each can see it, and the demonstrator has to turn his whole body in order to see it properly. This movement is her cue to zoom in quickly on the detail indicated so that the lecturer can look away from the camera without his face being seen.

92

Remembering What To Do Next

This early version of the hub and arc design described on page 12 helps the demonstrator to move easily from one display item to the next. The arrangement of the materials on his bench will help to remind him of the running order he has chosen. An additional reminder is the blackboard on which this lecturer has devised his own "lecture notes" in the form which suits him best. It serves him far better than someone else's script, laid out to suit a traditional broadcasting pattern; and it illustrates both the argument of

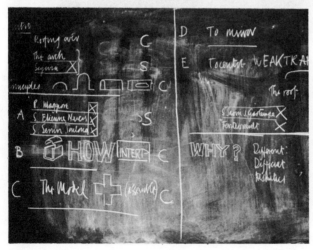

the lecture and the practical moves that must be made in order to introduce the model arch, then compare architectural samples (on slides), then return to demonstrate structural strength and weaknesses, and to develop from this the demonstration of a more complex model in which the underneath mirror frame is used to show plan as well as elevation; then more slides, and a summing-up spoken direct to camera.

In much the same way this lecturer has set out his notes to remind him not only of his subject but also of the demonstration points he feels most likely to forget. The camera lurks right beside the crib-notes.

95

This kind of support is a form of first-aid designed to help the newcomer who is used to relying on some form of lecture notes, or who wants to have something to fall back on whilst he gets used to this new experience. It is the operator's job to convince the newcomer that television's requirements are simple and specific, and to deploy the equipment so that it suits a lecturer's requirements instead of dragging him along at its chariot wheels. The first step is to make sure that all the materials required for demonstrations can be easily reached, without the need for the demonstrator to dive out of shot in search of them. Provide a "props table" or a shelf, behind and below the working surface the demonstrator uses and within his easy reach. Confine the area he has for each demonstration so that he cannot litter it with discarded tools and materials, and automatically disposes of them by placing them out of camera view. Alternatively, be generous with working surfaces, but establish clearly beforehand that separate sections are earmarked for each display. This leaves only two points to be made to the newcomer:

If you want to talk direct to the viewer look into the lens of *this camera* (indicating one camera which will always be available for "talking head" shots). When you want to show materials or equipment watch the effect of what you do on *this monitor.*

Storyboards

To give the newcomer the right support, the operator must have his own plan of campaign, a contingency plan in which he arranges his camera so that it is easy to follow whatever the demonstrator does. It helps to visualize the possibilities and to set them out in the form of a storyboard—a sequence of rough sketches that show, frame by frame, the views the audience will have on their screens. This is not necessarily an alternation of shots from one camera to another. The storyboard sequence on page 78 shows how a view may be totally changed as a camera pans from one subject to the next, or zooms from detail to a wide view of context. The main thing is to note down the different results that must be achieved on the screen, showing briefly how they are achieved by the operator, and what the demonstrator is likely to be saying or doing at the time.

This storyboard suits a very simple camera arrangement, in which Camera I does most of the work, going from one close shot to another, while Camera II provides cover for Camera I's changes, with a context view.

From such a beginning both the operator and the newcomer can develop and elaborate as they choose. The storyboard is primarily for the operator and any other television colleagues who may be involved. But it is also a useful way of checking in advance with a participant, and/or with whoever has commissioned the programme, on the nature and scope of the visuals and on the likely sequence of the exposition.

As a newcomer becomes used to the situation he can be coaxed into a deeper knowledge of the possibilities. After one or two successful appearances he may care to linger and be shown what

A I

Introduction:
HEAT TRANSFER

B II

The water's boiling—
here's the potato
(takes fork)

C I

(potato impaled)

D II *(as
for B)*

*(potato poised over
water)*

E I *(zoom-
ing in)*

You can see how
hot it is *(potato drops
in)*

F II *(as
before)*

It's inside now, get-
ting really hot

G I

Will it stay hot when I
lift it out?

H II *(as
before)*

Here it is, away from
the hot water *(remov-
ing it with fork)*

I II *(zoomed
in on
hand)*

(removes it from fork)
but—ooh! it's still hot
to handle!

the cameras can do in the way of overhead shots, underneath shots or zooming from context to detail. Do this with as little fuss as possible, and limit the do's and don'ts to the bare essentials. The newcomer can teach himself much faster once he gets used to using his monitor.

Later, having acquired confidence in the use of these basic facilities, some developments can be suggested: the working surface below the top mirror can be used to display objects, to compare plan views with elevation, to do simple animations—as well as merely demonstrating a craft skill. The worm's-eye view of an object or a process seen via the underneath mirror can also be used in over-drawing, animations, and silhouettes, or for the gels (acetate sheets) that once required an overhead projector. And the camera can pick up slides or a short film sequence (see page 44) just as easily as it can cope with photographs or charts displayed on a caption rack.

Rapport

All these devices add to the beginner's knowledge without requiring of him any special dexterity. But now he will need a little persuasion to try out the effect of over-drawing in order to label a piece of demonstration material. Let him begin simply by chalking a word or an arrow to add to the picture he sees on his monitor. From this he may soon be tempted to draw outlines and diagrams in register; still later he may use a simple animation, and subsequently try the effect of superimposing it on reality.

In this way we are enabling the outsider to feel his way into television, using it as much or as little as he chooses, and relying on the operator to do for him whatever he doubts his capacity to do for himself. With this increasing knowledge and confidence it becomes progressively easier to plan ahead, using a storyboard as a means of defining, for the operator's benefit, what the demonstrator intends to do, or requires to have done for him. In the process of formulating the storyboard, which may be on either's initiative, operator and presenter begin to anticipate each other's requirements, to establish a rapport on which they can if necessary improvise or elaborate as they choose.

Cueing

The moment when the presenter is most vulnerable is when he is about to begin. This is the point at which his uncertain grasp of the medium, and perhaps his suspicion of the operator, are most important. He is, he feels, dependent on this technician, and if he makes a fool of himself it will be the technician's fault, but no one else will know that.

To avoid such tension, take extra care to contrive an opening to the programme that does not expose the presenter unfairly. He wants to know "When am I on the air? How shall I know when to start?" Explain clearly, but casually, what signal he should look for to begin. Then make sure you allow sufficiently for his reaction time before cutting to him after cueing him (by cue light or gesture). Otherwise the viewers may be treated to that frightened-rabbit look as the presenter is caught waiting to begin, and then suddenly gets moving.

It may be better to choose an opening shot on a caption, cue the presenter to begin, and then as he says his first word cut or mix to the camera to which he is speaking.

Or one may leave the presenter to set the pace himself. Explain to him that the camera will open up on the apparatus or materials he is working on, or even on an empty set. Place a monitor where he can see what is happening. In his own time he may then look up from his work, or walk into shot, and start talking when he chooses.

6
Showing Things to Advantage

Display Areas

On every assignment you tackle, think first of the subject, and the *setting* that will do justice to it. Often this will be its own natural surroundings—the shop floor, the laboratory, the classroom, the gymnasium. Each provides a context in which the activity you are televising seems more authentic, and the participants feel more at home.

Sometimes, however, a studio setting is more convenient, perhaps because particular facilities are more easily available.

Whichever you choose, make sure that the most significant aspects of the subject matter get the best available lighting and the most revealing camera angles and magnifications; and that everything which might distract is got out of the way, or into obscurity.

When going on location, carry your own portable background (a roll of corrugated paper — see 97) to cover up distracting wall fixtures, as in this museum display room.

These museum specimens showed to best advantage when they were free-standing. In order to suggest comparisons in the woodcarver's style we wanted to look first at the ancestor figure in the

98

99

foreground (hung by a fine wire from a bar supported by the pole-cat shown on page 93). He was given a piece of blanket to rest his feet on, to lodge him firmly in the correct position. By setting both figures several feet away from the background they could be given their own lighting without distracting shadows being cast on the background.

Sometimes the most effective display stand for an object is a human being, as here, slowly lowering the wooden mask to reveal his own eyes staring above.

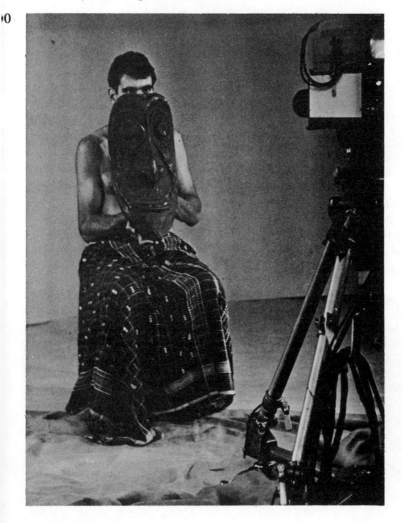

The tilt easel (shown in use with an overhead mirror in **64**) is here used to give variations in angle to a landscape model. Combination of the tilting movement on the easel and a panning and/or tilting movement on the camera can suggest changes in the inclination of the sun (represented by an Anglepoise stand lamp) or the effect of an aeroplane moving across the terrain.

101

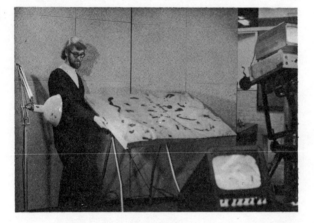

Fish are fascinating but elusive subjects. The insertion of an additional sheet of glass in the aquarium restricts the fish's movement towards and away from the camera, thus keeping it in focus, whilst leaving it freedom to move from side to side and up and down.

102

A shallow dish makes observation easier, although there is less opportunity to observe the way the fish moves up and down. Displayed beneath a small top mirror, shown here in the multipurpose display unit of the modular studio, movement can be seen and compared, either separately or both together, in both plan and elevation. Illumination is indirect, so there are fewer problems with the reflection of bright light from the water. The light intensity above can be varied by rotating a fader knob.

3

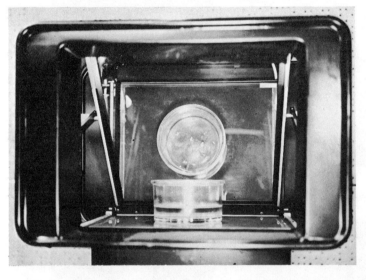

Alternatively, light the specimen from underneath, by projecting a beam of light on to the ground-glass screen on which the dish rests. The light source is the slide projector at the base of the display unit (see page 45), but without a slide in the gate. If desired the specimen can be shown in harsh silhouette, using the underneath light only, or a balance can be set by regulating the faders on the lighting above and below until a satisfactory effect is achieved. Decide this by experiment, comparing both the plan and the elevation views to see whether one is preferable to the other, or whether both should be shown together.

Ambient lighting shows glassware to good effect. It is sometimes useful to place the illumination behind the objects being shown. Glass and liquid show up against the evenly lit background —a frame of translucent plastic (which elsewhere does duty as a back-projection screen). The screen was manufactured as a ceiling panel, but is relatively cheap, and is lightweight. Hung between two trux it can be lit as shown by stand lamps, and/or clip lamps fixed to the edge of the trux (106).

104

105

96

Close Scrutiny

Close observation of objects and processes exaggerates every effect
—good or bad.

Make sure there are no distracting shadows. Care in lighting
need not mean complexity. Often an effect can be achieved with
very few light sources.

A shot zooming in on this majestic figure can be impressive
partly because the camera angle makes the idol look down on us,

97

partly because the lighting is harsh in order to accent the figure and the pillars, and leave the background dark and mysterious.

Two Anglepoise lamps on the close-up platform, and the job is done. Compare this with a more conventional lighting of the same model shown beneath the top mirror on page 55.

108

In most shots the zoom lens is augmented by a supplementary lens. Screw it on carefully, using both hands. This does the same

109

kind of job that a "portrait attachment" achieved for the old-fashioned Box-Brownie. The zoom lens continues to function on the same ratio (1 : 10 here) as before, but it can now be used on objects placed much nearer to it. Here are some typical distances:

Angénieux 1 : 10 zoom	Minimum distance without Lens	Minimum distance with Supp. Lens No. 1	Minimum No. 2	Minimum No. 3
(15-150 mm)	48 in. (1220 mm)	24 in. (610 mm)	17 in. (432 mm)	10 in. (254 mm)

Use of a supplementary lens imposes one further limitation. You cannot get sharp focus beyond a certain distance. So if we are to make the comparison complete we should also show these figures:

Angénieux 1 : 10 zoom	Maximum distance without Lens	Maximum distance with Supp. Lens No. 1	Maximum No. 2	Maximum No. 3
(15-150 mm)	Infinity	50 in. (1271 mm)	26 in. (661 mm)	12 in. (305 mm)

Every zoom lens should have one or more supplementary lenses available. Note down for yourself the maximum and minimum distances that your own lens combination allows.

It will be seen from these figures that a zoom lens equipped with an appropriate supplementary provides a useful range of shot for middle distances as well as for big close-ups.

The objects displayed on the working surface below the top

mirror are within the 50 in. maximum distance but supplementary lens No. 1 permits the camera to be brought in close to
the mirror, so as to command both a context view (via the mirror)
spanning most of the working surface, and a detailed view (direct)
of one of the lens boxes shown.

This situation can be exploited in order to cater for several
objects, varying in size, with the same zoom plus supplementary
lens. Here the camera operator is measuring the distance from the
back of the lens to the shell shown below the portable top mirror
on the table. She will bring her camera as close as possible, in
order to get maximum magnification. Using a supplementary No. 1
this means the distance must be at least 24 in., to show the shell,
front view, in sharp focus. Later in the programme there are other
objects she wants to show in plan, seen via the small top mirror.
But the distance to them is no more than 3 feet, via the mirror
surface. The caption on the rack can be placed further away, on
the same arc that includes both the model boat the demonstrator
manipulates and his face as he talks to camera. By placing demonstrator, boat and caption rack within the 50 in. maximum distance, each can be seen in mid-shot without the necessity to remove the supplementary lens. The only refocusing required is for
an overhead view in close-up of the objects beneath the mirror.

111

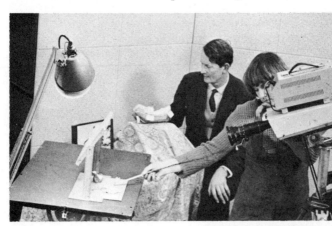

We can exploit the narrow limits of focus that use of a supplementary lens imposes. This crude landscape model becomes un-
cannily life-like on the screen if we place a figure in the foreground,
and *focus through.*

12

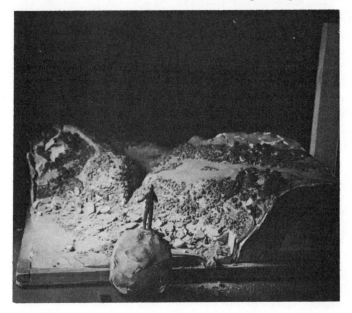

The shot opens with the lens set to show the area marked in dots on the photograph, with the figure in sharp focus, and the landscape he surveys a distant blur. Without changing camera angle, or the zoom of the lens, the focus ring is adjusted until the foreground figure becomes blurred and the landscape itself becomes sharp.

Another way to tackle such a shot, particularly where the quality of the model materials does not bear too close a scrutiny, is to focus for the half-way point between the foreground object and the background.

Lighting

The number and intensity of lights which the big television studios and Outside Broadcast units bring to bear sometimes deter the beginner. A prospective lecturer/demonstrator is soon put off if he is dazzled or frizzled by a massive concentration of lighting that bears down on him from a complex of stands and grids.

The operator is just as worried. On location, particularly, he must ration his demands on the local power sources, and the less gear he has to transport, erect and dismantle the better he is pleased.

Much of this complexity can be avoided. An elaborate studio lighting system is designed to illumine the whole of a large area, and often substantial sectors are lit to no purpose because no significant activity takes place there. The closed-circuit operator can afford to select a few areas in which he knows there will be activity, and to light those well. He is unlikely to be dealing with ambitious dramatization. His subjects will be working at bench, desk, display unit, or round-table discussion. If there are larger-scale activities these are likely to take place on the job, for example in a school, a gymnasium or a building site, where the addition of ambitious supplementary lighting would be unpopular and irrelevant.

The strategy to follow hinges on two requirements: the first is to make full use of natural or artificial lighting that is already available: the second is to place supplementary lighting inconspicuously, and where it can be concentrated on what matters most.

The placing of illumination is more important than the intensity of the illuminants. To give your subject impact it must stand out from its surroundings. This requires lighting that gives the effect of sunlight—a source of *key light* for the major illumination, and to balance it a lesser source of *filler light* to mitigate the harsher shadows. Placed at an angle with the subject, preferably slightly above as well as slightly to one side, the lamps are less likely to dazzle a newcomer, and they will begin to suggest a third dimension. The effect will still be inadequate, like a street scene when the sky has clouded over, until a third source is introduced to provide some *back lighting*.

"Key" and "filler" must between them ensure that the shadows that appear on the subject-matter are sufficient to give it shape and depth, but are not so harsh and intrusive as to obscure important detail, or produce a distracting pattern. The best modelling is usually achieved when these two types of light source are well separated and distinct in their effects. But avoid the sharp outlines of spot lighting by widening the angle of the lamps and reflectors to produce a "soft edge."

Back lighting can be even more diffused, and may be built up from several light sources—on a lighting batten for instance.

The best way to explore the possibilities of television lighting is to experiment with three low-power lamp sources, such as 150-watt lamps in adjustable Anglepoise fittings. Take any three-dimensional subject and light it, one lamp at a time, in order to see how the illuminants can be made to complement each other without neutralizing each other's effectiveness.

Then investigate the ways in which artificial lighting can be reinforced by daylight, and by light reflected from fabrics and wall surfaces. Notice how useful this diffused, *ambient* lighting can be. It is a natural "filler," it does not dazzle, nor overheat the participants. It helps to kill the hard shadows which can appear so dominant on a television screen that they distract attention. This effect of ambience can be contrived with the help of a diffusing material such as the screen shown on page 87. Alternatively lamps can be aimed at the upper parts of the wall, or at the ceiling, so that their light is bounced back on the subject matter instead of bearing on it direct.

With these differing requirements in mind, consider the variety of lightweight, easily adjustable lighting equipment that is available.

The lamp sources and fixtures shown here are suited to either a lighting grid—one or more horizontal poles suspended from the ceiling—or an upright "polecat," which can be sprung between floor and ceiling without the necessity for permanent fixtures. The

3

tungsten halogen or "quartz iodide" lamps (800-watt) shown at the top and bottom can be adjusted to give reflected light from ceiling or direct light downwards on to the subject—always provided that the tube containing the vapour which provides the light source is kept horizontal. The 500-watt Lowell light shown suspended from a grid bar is a very useful all-purpose fixture, so light that it can be held in the jamb of a door or in emergency taped temporarily to a girder or a wall bar. All three lamps are equipped with one or more *barn-doors*— flaps which can be adjusted to limit the area of light cast by the lamp.

This lightweight batten may be adjusted so that the lamps shine upwards on the ceiling, for ambient lighting, or downwards direct on the subject. Long hinges form the supports for the batten, with butterfly screws that can be tightened to keep it at whatever angle is required.

114

Four lamps are mounted on this flexible tube with a switch panel at the centre and a bracket that can slide up and down or pivot round its shaft (115).

The upright stand can double as the support for the mirror boom shown on page 51.

115

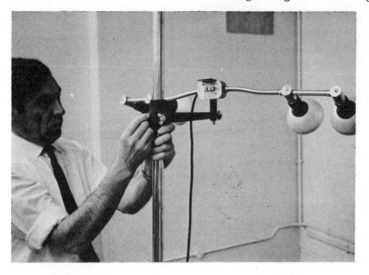

Daylight from the window, normal strip-lighting and a polecat provide enough ambient light with a few relatively low-power lamps placed close to the subject in order to pick out important detail and provide modelling.

16

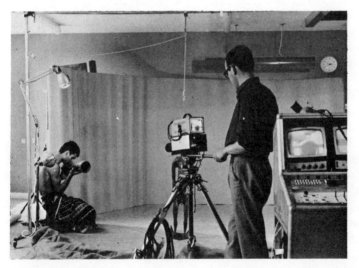

In this portable system (117), a free-standing, lightweight framework provides a very wide choice of lighting positions. The lamps shown

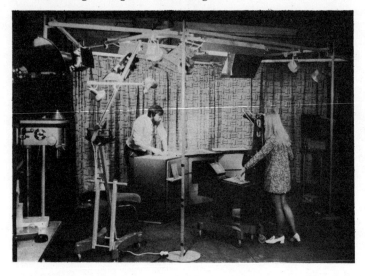

can be clipped to any part of the framework, which carries its own power supply within each shaft, safe from inadvertent contact by hand, but dispensing with the need for separate wiring for each lamp, since contact is made as the lamp fixture engages with the inside of the shaft. Since the lamps are placed fairly near the working areas that require to be lit, there is no need for the heavy, high-powered lamps which are often used in the more conventional studio, where light must be projected from far off. For the same reason, the total amount of heat given off by these lighting sources is modest, and therefore problems of cooling and ventilation are diminished. Notice that the framework makes possible the illumination of the surrounding area, as well as the area it encloses, if so desired. The enclosed area could be used for the display at close range of captions, objects, processes; whilst interviews, or a discussion, or the demonstration of a process or activity requiring more space, could take place outside the framework, with the mounted lights angled outwards to supplement what was already available from normal daylight or artificial lighting.

This interview situation (118) makes use of a big window looking northward on a partially enclosed area. There is no direct shaft of sunlight, so the effect of the window is to give evenly diffused overall lighting, which dominates the picture seen by the camera in the foreground. The interviewee's face would be in much too sharp a contrast between the window light on his right profile and

118

the shadow it leaves on his left profile, if it were not for the reflected light provided by the light-toned blanket draped over an easel to his left.

The two Anglepoise lamps, 150 watts each, besides providing additional "filler light," serve to illumine the face of the interviewer when she turns on her swivel chair towards the camera, as she introduces the programme. The area shown through the window, as background, is important. It would be fatal if there were no buildings outside, as the camera would then be exposed to the sky, and the contrast between the brightness of the sky and her face tones would be so extreme that she would appear almost in silhouette.

The use of a window position is valuable for several reasons. Whoever faces the camera can feel at home, in natural surroundings. The artificial light sources used as "filler" can be placed well away from his normal sight-line, so that they neither dazzle nor fry him. If the camera is working at close range to the subject, the vista out of doors will be sufficiently out of focus to be undistracting. Alternatively, working at medium range, an attractive background, of trees and bushes perhaps, provides something pleasant for the viewer's eye to linger on without diverting attention from the speaker.

As a variation, this use of a portable mirror stand (119) provides a useful reflection of daylight on to the right profile of the interviewer, and, when the camera requires it, a reflection of his face in

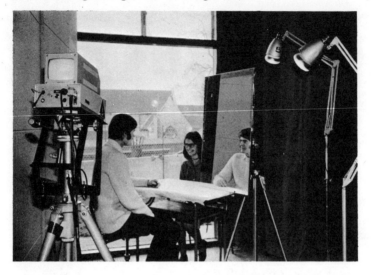

close-up. The interviewee sits where she can be seen by the camera
(i) in big close-up, or (ii) as a two-shot with the interviewer's head
and shoulders seen in the right foreground, or (iii) with both faces
shown side by side. The television camera shows the interviewer's
face on the left of the mirror, so that when panned left with the
lens zoomed out the mirror image will be almost cheek to cheek
with the face of the girl he is interviewing, and the effect is as if
the two faces were inlaid side by side.

The two Anglepoise lamp stands shown here are placed so as to
fill out the lighting on the girl's left profile, and to supplement the
light reflected by the mirror on the man's right profile.

Good lighting is achieved step by step, trying out one light
source at a time in order to gauge its effect, and judging it in
terms of the picture each camera reveals.

It is often preferable to work out a lighting pattern "on the
floor," using a monitor nearby to judge the effect of each adjustment one makes to the lamps within reach. If the switch panel is
elsewhere, in the control room for instance, arrange for a helper to
switch lamps on and off as required, whilst perhaps getting on
with other jobs in the intervals. If the switch panel is down on the
studio floor, you can do the lot yourself.

In either case adopt a foolproof procedure for identifying each
lamp and controlling it. Label each lamp boldly so that the number
is easily visible wherever you happen to be. The switch panel that

controls these light sources should be similarly labelled, with a
chart beside it showing the positions of the electricity sockets to
which the lamps are connected. This is equally important in the
traditional studio, equipped with a lighting grid from which the
lamps are seldom removed—and in a more flexible set-up where
many electricity points supply lightweight lamp units which can
be easily moved about. Each of these sources is labelled with a
letter of the alphabet, and a corresponding letter placed beside the
switch on the panel. As soon as lamp leads are plugged in, each
source of illumination becomes known by a number plus a letter—
"7B" means lamp 7 plugged into source B. (In the lighting set-up
shown in 47 and 117 there is no need for a separate panel, because
the switch for each lamp is within reach, either on the lamp or
clipped into the shaft which supports it.)

Aim at a balance of lighting that avoids the need for constant
readjustment of the lens diaphragm on each camera. Elsewhere, on
page 12, we have looked at a way in which camera operations can
be simplified by placing subject matter on an arc, so that there is
no need to refocus. The next stage in sophistication, whether the
lighting set-up is on a portable framework or a fixed studio grid, is
the insertion of independent faders for each light source. After the
position and angle of each lamp have been adjusted (with the fader
set at a little *less* than full lighting intensity), switch on all lamps
together, set and keep the lens aperture at a convenient stop—say
f7—and use the faders to vary the light intensity, one lamp at a
time until each camera's shots show up satisfactorily.

Viewing Conditions

The placing of *screen(s) for an audience* deserves more attention
than it often gets. Make sure that each is placed against an undistracting background (not silhouetted against bright light from
the window), and is not in the path of a shaft of sunlight or a
strong artificial light. Find a convenient height for each screen—
that is the lowest point at which each person can get a clear view
without a spectator's head in the way. Group the chairs in a horse-
shoe formation, making the most of the fact that a television
screen can be viewed by a wider sector than a cinema screen without noticeable distortion. It is better to have viewers seated in a
few long rows, all close to the screen, than in many short rows, because most television screens are small by comparison with the
cinema. The optimum distance from any screen, film or television,

can be roughly gauged by holding one's clenched fist at arm's length, and moving until the clenched fist appears to fill the screen. Check, and if necessary adjust, each viewing screen and loudspeaker BEFORE the audience enters.

The *replay of a video-tape* sounds the easiest and most obvious procedure. But disregard of a few simple precautions can easily undo much patient work.

Viewers are used to domestic television sets which show pictures that come on punctually and look right first time. The footage counters on some tape-recorders give only an approximate indication of the beginning of the chosen sequence. The error may grow if the tape itself stretches with frequent use. The situation is complicated when a series of extracts is to be replayed from different video-tapes. To ensure that the right beginning is made, supply yourself with additional empty spools, run each tape through privately to the correct point, and then (if the machine allows) lift the two spools off bodily and put them on one side until needed. When the sequence chosen for replay is completed, either switch off the screen or disconnect the video-tape-recorder so that the audience is not subjected to fireworks on the screen and a babble on the loudspeaker as the tape is rewound or replaced and a fresh sequence selected.

If you are in any doubt about replay quality, sit where you can see at least one audience screen from your control point. Either place your replay machine in front of and to one side of the screen, so that you can watch with the audience; or fit a cyclist's mirror to the screen where you can see it when seated out of their view.

7
Matching Audio-tape to Vision

Audio-tape is a useful way of increasing the value of visuals. A track, separately recorded, can be relayed to a video-tape-recorder at the same time that televised material is being recorded; or it can be added afterwards.

Commentary Tapes

Sometimes the audio-tape is kept separate from the video-tape so that it can be played alongside the video-tape replay for particular audiences. Observation of children's responses in a school class-room, or work study on the shop-floor, may require *different types of commentary for different audiences.* Or the same audience may see the video-tape twice, first with the local sound and speech that we recorded on the video-tape at the time; afterwards with the original track faded down, and an explanatory commentary tape played on the audio-tape-recorder. To play both tracks together might be confusing; with successive replays students can make their own assessment of the activities under observation, and subsequently compare these first impressions with the more penetrating comments of the expert.

If a piece of video-tape observation is particularly significant, there may be several different comments to be made on the same sequence of events. An educational psychologist, a tutor, a participant, might each have something distinctive to say. By recording each of these commentators separately on audio-tape, we can make many uses of the same video-taped material, without spoiling the sound-track originally recorded.

Audio-tape is easy to manipulate and to edit. So where the commentator is more at home giving spontaneous comments than in reading a script, commentaries can be recorded *ad lib*. This may also prove a time-saver. Suppose a craftsman whose skill has been observed on video-tape is persuaded to explain the significance of what he is doing. It may not be convenient for him to do this whilst he is demonstrating his skill: the equipment he is using may be too noisy, or he may be too preoccupied with the skill to think about explaining it as he goes along. But directly the video recording is completed, he can sit down and watch it replayed as often as he chooses. Each time he watches he is asked to comment, conversationally, with the audio-tape-recorder running. The television operator, or a tutor, sits beside him, holding the microphone as shown on page 28, and provides an occasional prompt if necessary: "This might be worth explaining," "How much pressure has to be exerted here?" Similarly, a class teacher, watching the replay of a classroom situation in which he was taking part, may have useful background information to give about individual children, and how their present interests or skills compare with an earlier stage in their development.

The first run of the audio-tape may catch valuable comments while they are still fresh. Or it may be no more than a useful rehearsal. Unless the first recording is outstandingly successful, it will probably be worth while repeating the video-replay-plus-audio recording, on the off-chance that the second take will be "even better." Do not erase the first audio-recording until after the second recording has been made. This is more than just a common-sense insurance. It can be used, quite powerfully, to exert psychological leverage on the performance of the commentator. Before any audio-recording is made, say: "Let's just try a commentary and see if it comes off first time. But don't worry if it doesn't. We've plenty of audio-tape, and it's no trouble to do it again."

After the first take, say: "Well, that's in the bag. But perhaps you might like to try it again just in case it comes off even better; remember whatever happens in the second run, we've already got the first tape to fall back on."

If the commentary is particularly important, and the commentator is uneven in his performance, you may, just possibly, end by picking out the best of both takes, and editing them together as described below.

"Counterpoint Sound"

A second use of audio-taped speech may be in the form of *un-scripted material recorded independently of the video-tape.* For instance a video-tape sequence might illustrate the surroundings or the activities of one individual, who at some quite different time is interviewed with an audio-tape-recorder about himself, his feelings, his ambitions. Sections taken from the interview are edited to provide an audio-tape which runs alongside the video-tape, so that the words provide a counterpoint, an oblique commentary on the visuals.

This counterpoint need not be confined to words. Music has been almost done to death in the cinema as a means of heightening interest, or of reinforcing emotion at a crucial point. There is greater potential in the use of evocative sound effects. These, like speech, can be used in either of two ways. They can be linked closely to the action on the video-tape, more or less synchronized with it. Or they may derive from something quite separate from the activities shown, which nevertheless relates to it. The bland voice of a radio announcer describing "another successful bombing mission in which many fires were started" counterpoints a visual sequence showing the situation from the viewpoint of a family within the burning city.

These devices depend on two basic audio-tape-editing techniques:

(i)   extracting material and rearranging and sometimes condensing it;

(ii)   spacing it out in order to bring the right sound alongside the appropriate visual when video-tape and audio-tape are played together.

Sorting out Audio Material

Save time by careful analysis of the original audio-tape recording from which it is intended to select material.

Set the footage counter at zero and spool forward, switching over to replay at normal speed at set intervals roughly equivalent to, say, one minute. Select a phrase or a sound at, or near, each of these intervals which will serve as a landmark, and note it down in the margin of your record sheet, leaving generous spaces in between. Then return to the start and play through, making much rougher notes in the body of the recording sheet as you replay. Keep an ear open for each "landmark" passage as it comes up, so that you can move down your sheet to the next section to continue your notes. In this way, you can codify the contents of the audio-tape in little more than the time it takes to replay the tape once.

If you have more time to spare, or the material is complex, begin by replaying the tape once, scribbling down on a blank sheet as many identifiable phrases or topics as you can manage to jot down in the time. Then re-spool and replay from start to finish, this time watching a clock, and adding timings in red to each phrase you are able to identify, in the re-run, from your jottings.

Once you have this bird's-eye view of the material on the tape you can identify the position of the passages you particularly want.

Extracting them requires care. It is so easy to snip out several sections, and inadvertently to confuse the section you want with the intervening material you intend to discard, or to pick up a short length of tape by the wrong end and splice it in backwards on.

The simplest way to avoid confusion where many extracts are required, some of them from different spools, is to *extract by dubbing*. This means using two audio-tape-recorders, one replaying the material you propose to extract, the other recording each section as you replay it, on a new tape. This is a good system, provided the link between the machines is a clean one, and no electrical distortion is introduced; and provided the extracts required do not involve editing in fine detail. It is still useful to insert short lengths of yellow spacing tape between the extracts as described below.

To extract by splicing, equip yourself with one spare reel, on which each extract will be spooled, once it is identified. Set up the original tape on your replay recorder with its own take-up reel. Spool forward to the beginning of your first extract, replay it at normal speed to identify it precisely, rewind to the start and cut. Remove the discard take-up reel, and substitute the extracts take-up reel. Wind on to this a standard length of red leader tape, say the equivalent of 20 seconds, and splice the end of this to the free end of the first extract. Spool on until you reach the end of this extract; cut; replace the extracts reel by the discard reel; spool on until you reach the beginning of the next extract; cut and exchange the discard reel for the extracts reel once again. Splice in a length of spacing tape—usually a distinctive colour, which makes it easy to spot the intervals between each extract (say 3 seconds), and mark out this length so that you can quickly measure up each piece of spacing as you require it. In this way, exchanging discard and extracts reels, you are gradually sifting out the extracts you require, and leaving the remainder tidily stored on the discard reel, where it can be found in the right order if it is required, or subsequently erased and re-used.

Splicing Audio-tape

Within each extract you may wish to make further cuts, to abridge, or to omit unwanted interruptions. With practice this can become a very fine art indeed. The first requirement is to identify the speech rhythms and inflections which are compatible with each other if two sections of tape are joined after the intervening passage has been omitted. It is wise *not* to cut the tape until you have listened carefully to the relevant sections at least twice. It is easy to lose track of these, however, unless you mark the tape with a yellow chinagraph pencil, on the shiny side. You can then spool back and forth, past the section you intend to omit, concentrating your attention on the way your first passage ends and your second passage begins. If in doubt, cut generously, allowing a little bit more on either side than you intend to use, and put your discarded section of tape carefully on one side, with a scrap of splicing tape attached to its leading end as identification. Then join the two sections that remain, and replay the few seconds on either side of the join. You should now be able to judge more accurately the rhythms of the two sections, and you can make your final cut so that it allows a natural breathing space between one phrase and the next. When you are quite satisfied with the results, get rid of your discarded section of tape so that it doesn't get mistaken for a piece you want to keep.

Make each cut by placing the tape shiny side up along a shallow groove, preferably in a metal editing block; cut diagonally; remove your discarded section, replace with the new take-up reel, whose free end should have been cut in the same diagonal so that the two ends match. Get them edge to edge, without any overlapping. Cut a piece of splicing tape about an inch and a half long and place it firmly and precisely to overlap the join. Press it down hard and evenly. It requires no time to set. Lift the spliced tape out of the groove and look carefully to ensure that there are no sticky surfaces, or uneven tape edges which could catch during replay.

Spacing out Audio Extracts to Match the Visuals

Perhaps the audio-tape extracts are to be fed into a television programme whilst it is being recorded. There might, for instance, be opening music, followed by live speech from the presenter, and later in the programme a telecine sequence, or a slide sequence, for which a commentary had been previously recorded on audio-tape. All that is then necessary is to set up audio-tape extract 1 (the opening music), replay it on cue as the programme opens, stop the audio-tape machine as the yellow spacer reaches the replay head, and stand by to resume the replay when the telecine or telejector sequence is due. This explains the importance of standardizing on the length of each yellow spacer on an extracts reel. Whoever re-plays the audio-tape gets used to judging the points at which to switch off or to depress the pause button, so that a predictable length of yellow spacing is left before the start of the next extract. In practice, a margin of a second is adequate, and the operator can therefore time the introduction of the new extract to match pre-cisely with the appropriate visual.

Another method suits programmes that require many short sound sequences, each one closely matched to the appropriate section of the video-tape.

Begin by timing the video-tape, noting precisely the point at which each sound extract should begin. Then fill in the period between the ending of the one sound extract and the start of the next by inserting a sufficient length of yellow spacing on the audio-tape to bring the next sound extract level with the visual it is to match. If both audio- and video-tapes are started off at the same time they can then be left to run. Provided your timings have been meticulous, picture and sound will synchronize.

8
What One Camera Can Do

Use of the single camera is a valuable technique in its own right, as well as a useful training method and a means of getting value for a small capital outlay in the early stages of an installation. Applications range from those in which the camera is left motionless as a recorder of the situation under study, to those in which it is on the hop like a flea in order to keep pace with rapidly changing circum-stances.

The Camera as a Means of Self-assessment

Athletes, preachers, actors, instructors, demonstrators, and many other kinds of craftsmen benefit when they have the opportunity to look critically at their own performance when there is no one else to see. The operator's job here is to set up a camera and micro-phone in a commanding position, with a video-tape-recorder ready for action nearby, explain the essential controls of the video-tape-recorder—then make himself scarce. The value of self-assessment is often the greater when there is no one else present, because often the performer is less inhibited, more inclined to let himself go, and more willing to replay his first effort, spot its weaknesses and repeat performance and replays until he can bring about the desired improvement. He may sometimes think he needs an on-looker at first, and this will be a useful opportunity for the operator to stand by and ensure that everything is working properly;.but the replay is usually best left as a solitary affair, and the sooner the performer can acquire confidence to manage the equipment on his own the freer he is likely to be in the use he makes of it.

The Camera as a Monitoring Device

Often the value of camera and video-tape-recorder in research work is that it records *without discriminating.* A single camera, placed strategically, will show everything that can be seen from that particular viewpoint. There are no cuts and zooms to "interpret" the situation—that is left to the research worker viewing the tape afterwards, and replaying it again and again until he has sucked it dry of information.

Applications include observation of traffic movement, archaeological finds and excavation techniques (using the battery-operated camera described on page 110), interviews and group discussions in which all the participants are to be observed all the time, animal and insect behaviour, and the study of stress symptoms in human physiology.

Here too the role of the operator is to make sure the subject is adequately lit and within the camera's focal range, and then to take a back seat. Probably he remains beside the camera to make sure it follows the action, but he must hold himself back from zooming in to close-up for fear he deprives the research worker of the context view that will enable him to interpret the significance of the detail.

The Camera as a Display Device

The most direct use of a single camera is to place it at a vantage point where it can show a process, or an object, to an audience which may well be in the same room but needs the magnification and the viewpoint of the camera to perceive the subject effectively.

The camera may be operated by the demonstrator, or by someone else he appoints, a student perhaps, and the amount of manipulation required is deliberately simplified. For example, he may wish to use mirrors on the lines described on page 49 to reveal a craft from the craftsman's viewpoint. The camera can be set so as to show the subject via a mirror suspended above it, and camera manipulation may require no more than adjustment of the zoom to go from detail to context.

The next step may be to compare the overhead, or plan view, with a front view to show the subject in elevation. This involves three simple adjustments, which should be practised beforehand: pan down from the mirror view to the direct view of the subject; change back the reverse line switch; refocus. The change in focus is the most time-consuming, but if there has been a preliminary check, the operator can gauge the extent of the focus ring adjustment, as described on page 116, and do it automatically. The result will be to effect a change of shot from overhead to front view with only a momentary interruption, and the audience will come to accept this in just the same way that they accept a frames change on the slide projector, during a normal lecture.

A single camera may also be used, via an overhead mirror, to show chalking by the demonstrator; or via the underneath mirror to show simple animations and drawings on the bench frame described on page 47. Here too it may be useful to change shot so as to show a front view of an object to which the diagramming refers.

Since both the overhead and the underneath view involve the use of reverse switches on the camera to the correct mirror image, it is wise to initiate the change from diagram shot to front elevation by switching back to normal, and immediately afterwards panning the camera to the new position. It is less disconcerting for the viewer to see a diagram change round from left to right or top to bottom than to see a three-dimensional object momentarily stand on its head.

A single camera will often suffice for simple interviews, in which the questioner's head and shoulders are shown in the foreground, in order to establish his relationship to the interviewee, who can be isolated in close-up simply by zooming in. 118 shows how the camera can also pick up a full-face view of the girl putting the questions when she spins round on her swivel chair. 119 on page 98 shows how one can go even further and put the faces of both participants side by side on the same shot.

All these devices are short cuts, which depend for their success on the operator taking the trouble to learn, and then stick to, a simple routine. Provided the routine is observed, the brief readjustments required will cause as little interruption as the pause to erase the writing on a chalkboard during the course of a lecture.

The Hand-held Camera

Some of the pioneer work in the use of a hand-held CCTV camera was done by Dr Aled William whilst at the Cardiff College of Education. He made good use of a camera tethered to his mains system by a lead long enough to permit the operator to move freely about a demonstration classroom observing the way small children responded to various teaching techniques and materials.

Battery-operation. The Japanese firms of Sony and Shibadan led the way with the introduction of the battery-operated manpack, a lightweight video-tape-recorder which can be slung from the shoulder, linked by a short cable with a hand-held camera. The camera is provided with a pistol grip which operates the video-tape-recorder. Alternatively a tripod can be used, and the recorder switched on from a button on its dashboard. The microphone is usually clipped to the top of the camera, and preset to a general-purpose level.

The simplest version of the manpack uses a reflex viewfinder instead of increasing the weight of the camera by the inclusion of a viewfinder monitor. Early versions required a mains-powered record/replay machine to replay the material, but more recent versions are providing a viewfinder monitor through which the portable machine can replay in order to check on programme picture quality.

The chief limitation of the hand-held camera is the lifetime of the batteries, which power both the video-tape-recorder and the camera and require frequent recharging with a mains charger supplied separately. Assuming there is one set of batteries in use, and a second in readiness, it is essential to plan shooting to allow time between assignments for the batteries to be recharged. A set of batteries which will keep the equipment running for say, 40 minutes, may require six hours for recharging. The other limitation is the length of the video-tape spool, which may be substantially less than on mains models, 20 minutes instead of 60 for example. It pays to double-check battery life before starting each assignment, and if there is any doubt to replace the battery set. The first sign that the battery is running down is that focus on the viewfinder begins to blur. To conserve the batteries (which are still being used up whilst the camera viewfinder is in use, even though the recorder is at "standby") choose each new viewpoint, and roughly check its picture composition before switching on and making fine adjustment with the help of the viewfinder. If there is then a delay, switch off again, but remain at the ready so that as soon as the chosen activity looks like beginning you can switch on, wait the necessary few seconds for warm-up, and then start recording.

The most flexible way to treat the television-camera-plus-video-tape-recorder is to use it as if it were a film camera. This means taking a series of shots, which together will add up to a programme, though they may have been recorded at intervals, so that whilst the camera was switched off the operator could move to a new position or a new subject. Because it can be carried anywhere and is independent of a mains supply the manpack offers a great variety of camera positions, achieved rapidly without fuss. An interviewee can be button-holed for quick comment as he continues his job, or stops momentarily on the street corner. In observation work one can stand on a table for overhead shots, kneel on the floor for a worm's-eye view, or shoot from a car or train window. One can move towards a subject, shooting the while, or move sideways to get a better angle when the subject moves out of direct view or is temporarily obscured by someone in the foreground.

Stability. The other side of the medal is that hand-held cameras are seldom held steady. Our brains accept the rapid changes of viewpoint that our own eyes convey to us as we walk about or look up and down. But when anything approaching this variety of movement is reproduced by a camera on the television screen the jerkiness becomes accentuated by the "frame" of reference the screen imposes. So the first requirement is to handle the camera as if it were a brimming goblet from which not a single drop should be spilled. This means hugging it to oneself, so that hands, wrists and forearms become clamps, and panning and tilting movement are effected by a deliberate movement of the whole body.

Continuity. The main value of the manpack, however, will probably continue to be its usefulness as a means of recording spontaneous activity that requires more frequent shifts in viewpoint and magnification, and less fuss, than use of a camera stand usually permits. The situation so closely resembles that in which a hand-held film camera is used that one can easily forget the difference in editing and cutting technique between film and television. The documentary film tradition has been to hew out the raw material on location, and create the programme in the cutting room. On this assumption there is no need to shoot the material in chronological order. One can film the long-shots and the mid-shots, and later on go back to reconstruct appropriate close-ups, knowing that provided the rules of continuity have been observed and the subject hasn't changed his coat between mid-shot and close-up, it will all fit together in the end. Even in film-making this approach is modified when high-speed newsreel work is required. To follow a swiftly moving situation—a Vietnam patrol, a strike meeting at the dock gates—and to get it out to viewers on time, means that there is very little opportunity for the cameraman to convey his intentions to the film editor in the cutting room, so he becomes more inclined to get the sequence right at his end, by shooting more or less "in continuity." The video-tape-recorder makes editing possible, simply by copying extracts to a new tape in the desired order; but this is time-consuming, and most users of portable television equipment have little time to spare. It is usually better to shoot the material as one would do if two cameras were available, that is to say changing viewpoints in chronological order but switching off for a quick change to a new position. To do this effectively requires careful thought beforehand to establish in one's own mind not so much the eventual shape of the programme, which the normal sequence of events will dictate, as the kind of emphasis the camera is required to give.

A problem in observation. Take the Primary teacher at work with his class, as he launches them off on a project topic. The recording is designed for student teachers, who want to see how an experienced teacher stimulates interest, provides information, organizes activity and gives individual guidance. This particular teacher starts off with the children in rows in front of him, and he begins to talk, and to chalk on the blackboard just behind him. As he develops his subject he asks questions, and invites suggestions from the children, and he jots down on the blackboard the ideas that arise.

There comes a moment when the flashpoint is reached, enthusiasm is generated, and the class can burst into activity. There is a general reorganization, as children split up into groups to tackle different assignments. Some are looking things up in reference books from the class bookshelf; others are collecting materials to paint, or model, or cut out. Others are discussing their plans. The teacher is no longer at the head of the class, easily located at a fixed point. He is on the move, helping to check out materials, advising, explaining to individuals and groups.

Gradually the groups begin to settle down, and different pieces of work take shape. Some children are moving about collecting information which they will tabulate on graphs. Others are making drawings, or cutting pictures from magazines to make up a collage on frieze paper that will go the length of one wall. Two children are collaborating in the making of a scrap book "magazine." The rest are at desks or round small tables writing, drawing, arguing.

At a much later stage, these scattered activities are brought together as the teacher, returning to the blackboard, gets individuals and groups to report back on the progress they have been making to the rest of the class. At this point he may be holding up charts, pictures or models that are being made so that everyone else can see them.

A sequence of viewpoints. Such a variety of activity requires a nimble camera operator to do it justice. The operator might begin the recording by standing near the back of the class with the camera focused on a caption card propped up in the foreground beside a clock. This way he labels the location and the time, and then zooms out to a wide shot, gradually adjusting focus as he does so, to show as much of the class as his camera can command, and in the distance the teacher and his blackboard. Then he zooms in to show the teacher, head and shoulders, and to pan off to the blackboard as ideas are jotted down on it. At any moment the teacher's relationship to the rest of the class can be quickly suggested by zooming out, to show how many hands are up and waving in response to his questions.

A little of this view from the back of the class goes a long way, however. The opening shot has labelled the subject, established the situation, introduced the teacher and his blackboard, and suggested the lines on which he is going to work. Whilst he continues there is time to switch off and move half-way up on one side of the class where there is a closer view, oblique but still legible, of the blackboard, and a rapid panning movement will show the class, with the rear section visible in full face or in profile. Their interest in the subject matter is awakening, so they take less notice of the operator, who is still relatively inconspicuous, in the wings as it were. The advantage of this position is that the shift of emphasis from teacher to class and back can be achieved by a zooming movement. The teacher is in close-up; as we zoom out from his face we can see the blackboard behind him, and as the zoom continues the heads of the front row of children appear. We can now pan downwards and begin to pan across the class, so that the children now fill the screen. At first it is only the backs of their heads, but as the panning movement continues, still fully zoomed out, the profiles of the row level with the camera operator come into view, and then, full face, those at the back of the class. The distance from camera to teacher is roughly the same as that from camera to back of class so focus requires no readjustment to zoom in now on these faces at the back, which are the indices of the

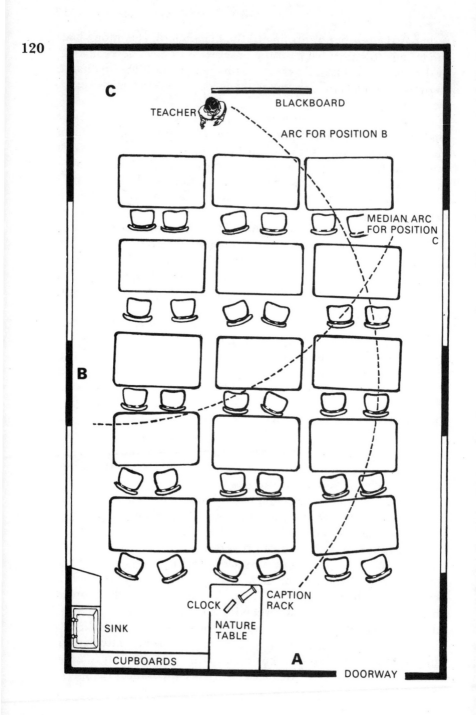

C

TEACHER

BLACKBOARD

ARC FOR POSITION B

MEDIAN ARC
FOR POSITION
C

B

CLOCK

CAPTION
RACK

SINK

NATURE
TABLE

CUPBOARDS

A

DOORWAY

teacher's success.

Technical requirements. The opening shot (from *A* on the diagram) was tricky because it needed a change in focus from the foreground caption to the distant blackboard. Before the shot began, the operator looked through his viewfinder at the blackboard, zoomed in as close as possible, and adjusted focus. Then he panned his camera down until the caption card was sharp and the blackboard blurred. He practised this change two or three times until his hand was used to the movement of the lens focus ring, and he could make the change automatically. This is easier to do than to write about; one simply notes that when grasping the focus ring, one's thumb is, let us say, at 12 o'clock for the caption card, and at 5 o'clock for the blackboard. Since the depth of focus improves as one zooms out, the operation may be broken up into three easy stages:

> Opening shot, zoomed in on caption card.
> Zoom out and pan up to reveal class and distant teacher.
> Readjust focus whilst zoomed out to approximate sharp focus on teacher.
> Zoom in on teacher and blackboard, and when fully zoomed in make any fine focus adjustment still required.

Next comes shot *B* from halfway up the classroom. This may require special attention to lighting contrasts. There may be windows along the length of the classroom on both sides, and if they are shown in shot the brightness of the light outside will contrast sharply with the interior of the classroom. If only one wall has windows in it, the choice is simple: the camera operator chooses that wall and is able to survey the classroom with the window light behind him. Otherwise he must practise before making the shot so that when he zooms and pans as he shifts from teacher to class he keeps the camera tilted down so that the area it commands stops short at the lower edge of the windows. If he has to include children whose heads and shoulders are silhouetted against the light, he should also practise a quick adjustment of diaphragm to compensate for the change in light intensity. Focusing problems at this stage are fewer because the shot follows the arc shown on the diagram; but it pays to make a preliminary check on the focus adjustment required in case it should be necessary to zoom in on children in the foreground. The teacher's introduction has now generated a good deal of enthusiasm, and his blackboard is filling up with suggestions. It may be useful to complete the second shot by returning to the blackboard for a careful look.

This is a natural stopping point and an opportunity to move to the far corner of the classroom at *C* where the camera looks past the teacher (in the foreground) to the children. The operator is now conspicuous, but the novelty of his presence has lessened and the children's absorption in the lesson is increasing; so he can afford to open with a wide shot that spans as many of the class as possible, and then zoom in, losing the blurred foreground outline of the teacher, to concentrate on one face at a time.

Here again there are the twin problems of light contrast and focal distance. Before he begins to shoot, the operator makes a quick check to see what is the most convenient and rewarding arc on which to set focus. The diagram shows his choice. But this time he wants to be able to look at children in the foreground and in the background as well, so he practises making focus changes on either side of this median position. The nearest is 9 o'clock, the farthest is 4 o'clock. He has also checked on problems of light contrast posed by patches of sunlight, or marked differences in the tones of clothing. That small boy in the whiter-than-white shirt is going to be a problem, and so is the little girl with the dark hair in the shadow.

This third vantage point should prove useful as the class begins to break up into groups, and the teacher moves among them. There may be materials to be fetched from shelves or wall cupboards, and spread out on tables or on the floor. The shot can conclude on this general movement by zooming in as one group forms, to show them as they begin to get to work.

From now on the sequences are shorter, as the camera operator moves around the classroom to gain different vantage points for observation of particular activities. His presence is less and less noticed, because the novelty has worn off, and the children are getting to work on their special interests. But they may still become aware if he makes too much fuss in positioning himself and aligning the camera in the early stages of this activity. So he studiously ignores what is happening under his nose, and appears to concentrate his attention in the opposite direction. What he is actually doing, however, is pointing his camera at an object which he estimates is about the same distance away as the group he intends to shoot, and adjusting focus on this object so that when he turns to survey the chosen activity focus is already sharp and no further adjustment is required.

He has then to decide how to relate detail to context, by zooming in or out. At first he may want to establish the group as a whole and starts with a wide shot, afterward narrowing the shot to a close-up of a face, a pair of hands, a model or a chart in preparation. Later, as the work develops, and the children are thoroughly used to the close presence of the operator, he may begin zoomed in to a big close-up of a piece of handwork, which required extra fine focusing to get the details sharp, and then widening the shot to reveal those who are involved.

As the work progresses, the operator can take up more conspicuous positions, and may still go virtually unnoticed. He can stand on a chair, or even a table, and look downwards on several activities across which he can pan in yet another arc—one nearer the vertical than the horizontal plane. In this way he can pass from detail to context on one group, shift onwards to the context of another group and zoom in to concentrate upon its detail. Or he can zoom out and pan upwards to reveal many groups at work, and the teacher moving among them.

As the lesson draws to its close and the teacher begins to pull the class's activities together, the operator may find himself back again on the sidelines, at point *B*, where he can zoom in on the teacher and the things he exhibits to the class, and pan round on a wide arc which will reach many of the participants and keep them in sharp focus. As the class disperses he can complete the sequence by zooming in on one of the exhibits pinned up on an easel, or displayed on a nearby table.

As an afterthought, when the children have all left, he may invite the teacher to explain the significance of what has happened, to point out individual pieces of work and comment on the way the children responded. This too can be done on an arc, with the teacher moving around the class from point to point, and the operator, his back to the windows, panning with him as he moves.

9
Location Work

This poses human as well as technical problems. The first person to consider is yourself: what help can you obtain in transporting and setting up equipment? How much time is available? a dusty answer to these questions would suggest reliance on the one-camera techniques already described.

If a more ambitious treatment is in order you have then to consider rather carefully whether to fit into the situation *as it is* or to adapt the situation to suit the resources you bring to bear. The two approaches are described below.

Fitting in to What You Find

A factory, laboratory, nursery school, or gymnasium is visited because the activity can best be observed in natural surroundings, and it may require apparatus or facilities that could not be transported elsewhere. Television enters as an intruder, eyed askance perhaps because it is assumed that the paraphernalia associated in the public mind with Outside Broadcasts will disrupt the normal pattern. People expect, and perhaps fear, an invasion of technicians, script girls, and "toffee-nosed producers." If it can be shown that the operation is relatively undemanding, attitudes will change quite rapidly, and the atmosphere of strain is dissipated.

To make the most of what time is available requires careful reconnaissance, and a systematic testing of camera angles and positions. A preliminary visit can be unobtrusive but very rewarding. No one minds a visitor who stays in the shadows, and quietly watches an activity until he has isolated the significant details, and established in his own mind how these details can best be revealed by the camera. Given this forethought, much time-wasting manœuvring of equipment can be avoided. It pays to watch an activity for long enough to notice how people go about their normal work. Otherwise, a machinist may be standing by whilst the television operator examines the situation, the camera may be positioned and the apparatus lit, only for the cameraman to find that the craftsman's shoulder obscures or over-shadows an essential detail once he resumes work.

A well-lit factory, with skylights as well as windows may re-quire no additional lighting. This operator has clipped a mirror boom to the top of his camera station, so that one camera can cover both front and overhead views of the lathe operator.

21

Here the work is less well lit, and so a bar with a couple of 150-watt lamps has been clipped to the camera station to improve matters. The mirror boom is now secured to its own independent stand.

122

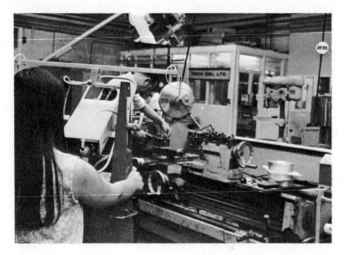

This craftsman is observed by an unmanned camera. No special lighting is required because the workshop itself has good general lighting and the subject matter—the glass-blower's flame and the tools he manipulates—shows up well. The camera angle can be exactly determined because the craftsman's work is precisely located. Once the diaphragm has been adjusted to suit the blue

123

flame, there is little more to be done. A supplementary lens shows all the detail required. If it was intended to zoom in even closer, the operator would move across from Camera II, which gives the context shot shown in 123, and zoom in on Camera I for the big close-up, adjusting the diaphragm as he does.

(a)

Often the observation of fine detail requires the cameraman to see the context of the detail his camera shows. The craftsman, shown here at work on an X-ray tube, can best be seen by the camera peering down across his right shoulder. The operator has placed himself at the side of the camera station, and is using the monitor incorporated in the camera plinth so that he can adjust focus and zoom without difficulty, and at the same time cast an eye direct on the craftsman's hand and his apparatus.

Here the camera station has been reduced to half its size so as to provide a low-angle view of the mechanic at work on the racing car. The boom mirror is there to provide a top shot when necessary.

125

A more sophisticated approach was involved in this observation of computerized control of a machine process. Camera III, in the foreground, is in the gallery that runs beside an office suite. It can serve three purposes. Panned down in its present position it shows

126

caption material. Panned right it views office activity, including the work of the programmer, through a window in the office partition wall. Panned left it gives a context view of the equipment under study.

Except when the technician comes to check the work, we are only interested in the operation of the machine tools themselves. There is no question of "a craftsman's viewpoint." Camera I shows a front view of the drill, and can be zoomed out to show the computer in three-quarter profile in the left foreground. Camera II provides covering shots.

27

The whole set-up can be controlled on the shop floor if required by the video-tape-operator, reaching across to switch or mix as directed. Monitors (one for each of the three cameras, plus a programme monitor) have been placed on a table, with the portable vision-mixer beside it. Compare this with pages 23—4.

8

This lecture room situation poses a similar problem. Neither the audience nor the lecturer wants television to interfere with the proceedings by distracting attention or introducing lighting units that get in the way. Fortunately the room is well lit, with a light ceiling that reflects and diffuses the illumination provided by the hanging lamps. Camera and video-tape-recorder are there—but unnoticed.

129

Here they are, three rows from the front.

130

The lecturer can be shown in medium close shot, or the camera can pan with him to follow his work on the blackboard.

131

The same sort of unobtrusiveness is achieved here in a lecture-demonstration on microscopy. Students can see on the screens suspended from the ceiling what the lecturer observes through his own microscope, or sketches out on the writing frame. (See page 67 for a detailed arrangement of the equipment in a studio setting.)

32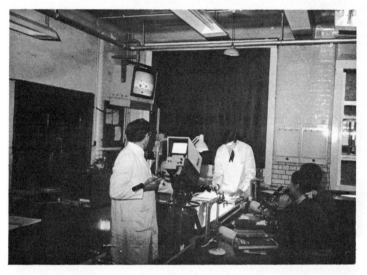

After demonstrating each point the lecturer refers students to their own microscopes to verify his findings with their own slides.

33

Although he is far away in one corner of a big laboratory, his voice can easily be heard through the loudspeakers, and the atmosphere of the whole undertaking is informal and purposeful.

134

Setting up to Suit Yourself

It is tempting to think first of camera angles but the major problem on location is often to get the best acoustic. Usually the best area will be a corner of the room where nearby walls reflect the wanted sound without a delayed echo. Compare the surfaces to find which absorbs sound best. Wallpaper is better than shiny plaster. Curtains and soft furniture help to limit reverberation.

Next, consider camera positions in relation to the background. There may be bits of equipment, or piles of half-used materials, or wall calendars, which will steal the picture; get rid of them tactfully, or rearrange them to reduce the number of loose ends.

Define the area in which the activity to be televised can conveniently take place. Site Camera I so as to command a satisfactory context view, with the widest angle of your lens comfortably spanning the action. This need not mean placing the camera so far away that it commands the entire working area in one shot, unless it is necessary at any point in the demonstration for everything to be seen at once. It is more likely that the demonstrator will be moving from one point to another, and that at any given time we shall not need to see more than a mid-shot of his face, the upper part of his body, his hands, and the materials or equipment he is using. Place the microphone where it will not impede him.

Camera I should command any of these activities simply by panning in the horizontal plane. It will of course reduce focusing problems if the activities occur roughly on an arc with the camera, and it may be possible to arrange this in consultation with the demonstrator; but it is important not to cramp his style by making him move in an unaccustomed way.

At this point you may have to give further consideration to distracting backgrounds. There may be table or bench surfaces that loom larger than life when seen in close-up. A piece of essential equipment featured at one stage in the demonstration must not be allowed to intrude in later shots that have nothing to do with it. A minor adjustment in camera position and height may ensure an undistracted view of what we want to see. Alternatively, interpose your own background material. An empty caption rack placed on a bench may be used as the display area in front of which the demonstrator is asked to show anything which has to be seen in detail. If the subject-matter is better shown on a horizontal surface cover it with a piece of fabric—perhaps no more than a blotter or a table-mat, on which the demonstrator lays tools or materials. Two advantages accrue: (a) you can choose the tone of the background material to contrast well with the subject detail, and deploy your light to show it to advantage; (b) provided these close-up demonstration areas are convenient to him, the demonstrator will be more inclined to locate his activity in the precise spot that suits your cameras best.

You may require a mirror view, either from overhead, or from one side, which can be conveniently shot by Camera I. Place the mirror and check that it does not intrude in any of Camera I's context views and does not get in the way of the demonstrator.

Now take stock. How easily can Camera I be realigned for each of the shots so far attempted? Is there a convenient pattern of operation you can adopt? Are focal distances the same for all the shots on each camera? If not, note down and practise the changes in focus that you may have to make in a hurry. What aspects of the action are inadequately revealed? This brings us to Camera II.

Camera II can be used to specialize. It should get the tricky shots which Camera I cannot command; and it may also give cover to Camera I when attention must be concentrated on one particular activity which must be seen from different angles, or in different magnifications. If efforts to eliminate distracting background in some of Camera I's shots have been unsuccessful, Camera II may command a viewpoint which includes less irrelevant clutter. Having made these decisions follow out the same procedure for Camera II as you adopted for Camera I: allow for the probable movements of the demonstrator; get as close as the best context shots will allow; suit height and angle to get the best from close-ups.

There may now be need for a minor readjustment of the microphone position, to prevent it from coming into close-ups. Small changes in position will not affect the sound quality, so long as the microphone is angled towards the subject and away from reflecting surfaces and irrelevant background noise. Now that we know the areas that will be in vision, it may be possible to introduce additional material to improve the acoustic by breaking up the sound, or absorbing it. This may be no more than a blanket draped over a draughtsman's easel or a blackboard, or the curtained screens sometimes used to make a playhouse in the Primary classroom. If its appearance is undistracting, it may form part of the shot, if not it can be placed just out-of-vision.

Now for lighting. Consider first what contribution the natural lighting already provides. Switch on artificial lighting to supplement what comes in through the window. Beware of patches of sunlight that move around and catch you unawares with distracting highlights and harsh shadows that intrude halfway through the recording. If there is any danger of this happening, put up some translucent material—butter muslin, or a sheet of tissue Sellotaped to the window—to diffuse the sunlight if and when it comes. Check carefully the effect of artificial lighting on your shots, particularly the close-ups involving shiny tools or materials that might reflect a ceiling striplight straight into the eye of the camera. Identify the source of light that does the damage, and switch it off or diffuse it. Consider whether the balance of lighting in the room can be improved by introducing a light-reflecting surface, such as a white chalkboard.

Having made the most of what you already have in position, you can now, at last, introduce supplementary lighting; the less the better because it may dazzle and distract the participants. Build up your lighting pattern one lamp at a time. Begin with your most powerful lamp, which will be used to balance the main source of natural lighting in the room. If the light from the window is strong, it can be regarded as the "key" source, and your strongest lamp will be the "filler," lighting up the darker areas of activity, and placed so as to illumine the aspect that the window lighting leaves in shadow. If, however, the working area is distant from the windows, or there is insufficient daylight, your own supplementary lamp will become the "key" light, and it need no longer be placed to complement the lighting from the windows.

Check through the main shots to see what problems arise. The cameras, acoustic screening, mirror and microphone are all potential sources of unwanted shadows. Subject-matter may include shiny metal or glass which produces awkward reflections. Lighting may spill over on to the viewfinder screens of the camera, or on the demonstrator's monitor. Get someone to stand in as participant to find whether there are awkward shadows as he manipulates material, and how dazzling the lighting seems to him.

Decide whether to increase the ambient lighting, by directing supplementary lamps at light-reflecting surfaces *outside* the camera's vision—the ceiling, the upper part of the walls, white surfaces of chalkboard or acoustic baffle. Can the area to be displayed be better lit by fewer lamps placed closer in? Will an additional Anglepoise lamp help to accent a significant detail, or to reach something awkwardly placed?

Having got your dispositions right, note them down for future reference, in case you want to come back to this place for another recording, or have to contrive a similar situation elsewhere. It is worth jotting down measurements on a sketch floor-plan, in particular the heights of lamps and cameras from the ground, and their distances from the subject matter at which they are aimed. If you have a still camera, take one or two general shots of the whole set-up, as reminders of the way the equipment was deployed; with a close-up of the subject-matter as seen by the television camera, both in context and in detail. Contact prints will be useful on file to help identify the recording.

Plug in the video-tape-recorder and make a short test recording. First the test card (on a caption rack which is easily commanded by whichever camera will not be used for the first shot in the programme, and conveniently lit by the lighting already in position). Then a brief sequence of the essential shots, and a sound test, spoken preferably by the participant in his habitual voice (not a repetition of a meaningless phrase—"Can you hear me, can you hear me, testing, testing"). Replay, make any readjustments in lighting, or in acoustic that the tape reveals are necessary, then spool back just short of the test card recording, which will serve both as a guarantee that the tape was properly tested before the recording proper began, and as an identification for this particular assignment. Check the footage indicator is at zero. Enter subject and date on the index card within the video-tape carton. Check any programme captions required in the programme, and see that they are in their correct sequence behind the test card.

Move around the activity area looking at the situation from the demonstrator's point of view. Check that the tools, pointers and materials needed in the demonstration are ready and within easy reach. Tidy away or cover up loose cable wherever it might otherwise trip him up.

Now look critically, from a layman's viewpoint, at your own side of the exercise. The worst thing you can do is to give the impression that television is a complex tangle of bits and pieces and that television operators are never ready for action on time. Stow away every piece of your own gear that is not in use. Check that supplementary lenses, if required, are easily to hand, and safely placed. Close doors or windows that might admit unwanted background noise, warn the switchboard not to put through calls on the telephone once the recording begins. When all these preparations are completed, then, and *only then*—tell the participants that you are ready. Nothing sets you and them up better than to be as good as your word.

10

Keeping Track

Storage

The rack shown here contains an assortment of materials for the
making of models, and the preparation of captions and charts.

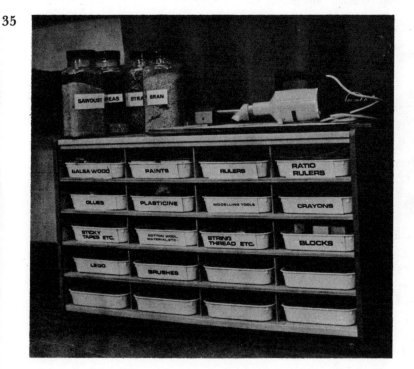

Model-making opens many possibilities and can prove a relatively easy craft to master, provided that the model-maker is prepared to try out each material he intends to use under the camera, in order to judge tone and texture. Shown in close-up, with careful adjustment of two or three light sources such as Anglepoise lamps, sophisticated effects can be achieved with cheap materials.

Slides will normally be kept in their own separate storage cabinets as part of an audio-visual resources store; but it is useful to keep one temporary storage space for special slide sequences used on the automatic slide projector. The smaller portable projectors are available for quick inspection when slides are being sorted out or selected.

There should be a range of *cards* of different tones and textures that can be used in the animation work described on pages 59–61, either to make foreground "scenery" or to provide untextured backgrounds on which the symbol to be animated can be drawn or pasted.

A sheaf of Letraset *types* chosen to suit size 12 × 9 in. captions will help to set a "house style."

136

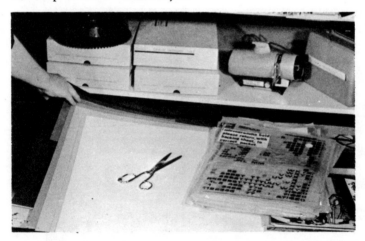

Logging

Most of us hate filling in forms, when they merely provide someone else—the tax inspector perhaps—with our hard-won information. What follow are suggestions for the reminder-lists that an operator needs if he is to keep abreast of his commitments and keep track of his equipment. Until you find yourself out on an important job and lacking an essential piece of equipment or, worse still, back from the job with an indispensable lead left behind, you will happily disregard all such paperwork. Later on, perhaps, you may feel it has its value. After all, every detail listed is exclusively for your own benefit.

First, get your briefing straight, and make sure that whoever commissions the job has thought out its implications. Experience suggests that the best way to use the *Facilities Booking Sheet* is to fill it in yourself as you check relevant sections with your client (page 134).

Next, check over the items of equipment the job requires. Make your own *Equipment Check List* (page 136). It can be used as a reminder of studio requirements as well as for Location assignments.

Note and number every recording you make, on video-tape or audio-tape, as you make it, by double entry: in your *work diary* and on an index card kept preferably inside the carton of recording tape. In this way you can quickly assess the volume of recording work your equipment is undertaking from one period to the next; and you have a double check on the subject matter each tape contains.

Finally, keep track of your equipment's reliability by entering every item you buy or make up, on a card index that allows space to *log faults* as often as they occur, with dates on which local repairs were made, or the equipment returned to the manufacturer for servicing or replacement. This log is the most important weapon you have in the battle for quality, because it can reveal a pattern of intermittent faults from which an engineer may be able to diagnose and correct a more fundamental weakness in the system.

TELEVISION FACILITIES BOOKING SHEET

Name.. Dept.......................Office Tel Ext...............
and/or Home Tel...............................

1. **What do you want us to televise for you?**

    A. OBSERVATION OF UNREHEARSED ACTIVITY (e.g. interview practice, classroom activity, movement, improvised drama, role play, teaching practice, group dynamics)

    On location *outside* the premises? ....

    On location *inside* the premises? ....

    In the studio? ....

    Who is the local contact (head
    teacher, departmental head, etc.)?...

    (address and telephone number) ....

    Will you be available to join us on the reconnaissance? ....

    Choice of dates and times (Friday mornings, at least two weeks ahead, preferred)..

    What is the deadline by which you want the job done? ....

    Is there only one *precise* date on which the activity can be observed? ....
    ....

    When would you be available to view material:
    (i) *during the visit* (e.g. lunch-break) ....
        (to check on how we are interpreting your brief)?

    (ii) *afterwards* (to select the section(s) most suitable for replay to your audience), and/or to record an audio-taped commentary? ....

    B. RECORDING AND/OR RELAYING A SPECIALIST DEMONSTRATING A SKILL (e.g. lectures, craft processes, laboratory techniques, etc.)

    Where? ....

    When? ....

    How long will room be available beforehand for rigging? ....
    (one hour beforehand preferred)

    Is permission required for right to subsequent replay? ....

    Will supporting material be used? ....
    (e.g. slides, charts, film sequences)

    If so, what? ....

    When can it be made available for check over beforehand? ....
    ....

    C. RECORDING YOUR OWN PROGRAMME IN THE STUDIO

    Working title of programme ....

    Target audience ....

    Do you require help in preparing the visuals? ....

    When can you bring visuals, storyboard, and props plot for checking over (at least 2 weeks before recording date)? ....

    What are your preferred rehearsal and recording dates? ....

    D. RELAY AND/OR RECORDING OF BROADCAST MATERIAL

    Title ....

    Time(s) and date(s) of broadcast(s)—including repeats ....
    ....

    Channel ....

    If the programme is *not* scheduled as a Schools or Further Education broadcast, special permission to record is required. Have you got this? ....

2. Relays

Do you want the material relayed "live" to an audience in a nearby room? ....

If so, when? ....

How many people? ....

Which room(s) have you booked? ....

3. Replays

When do you want to show recorded material to an audience? ....

Which room(s) have you booked? ....

4. Retention

Are you hoping to retain a section of the recording for further use at a later date?
....

If so,

(a) When could you view the tape to choose extract(s) and make notes for the catalogue? ....

(b) Can you arrange, if the tape proves suitable, for the cost of retention to be transferred to your department? ....

A SPECIMEN LOCATION EQUIPMENT CHECK LIST

Assignment No...................... Date due out........................ Date due for return...............

Van leaves base .... hours   Due at location.....................hours

Address... Telephone No........................

Van access via.. Parking..................................

Contact on reaching location..

Power sources available: 15A .... 13A(Flat) .... 13A(Round) .... 5A ....

Total rating ....

	R¹	L	R²		R¹	L	R²
Video-tapes (to record) (quantity) Video-tapes (for replay) (Cat Nos) Video-tape Recorder(s) (makes)				Audio-tapes (to record) (quantity) Audio-tapes (for replay) (Cat Nos) Audio-tape Recorder(s) (makes)			
Camera I (make) Camera II (remote) (make) Camera III (make) Camera Mountings I III				Radio Mic A Radio Mic B Moving Coil Mic Dynamic Mic Multi-mics (4 boxes+4 cables) set A set B Parabolic reflector Headphones Mic Cables Sound mixer (in v/m trolley) Auxiliary mixer (for multi-mics)			
Remote control pan/tilt head Remote control lens assembly Remote control control panel Camera wheel bases I III							
LENSES 10 : 1 and supplementaries 4 : 1 and supplementaries							
Vision-mixer trolley				**LIGHTING** (lamps and fixtures) Mini-quartz Verti-lites 1500W floor Anglepoise 150W Clip-on 150W T clamps Polecats (long) Polecats (short) Light stands			
Monitor(s) 9" Monitor(s) 23"							
CABLES Mains to control HD Control to VTR (long) Control to VTR (short) Control to Cam I Control to Cam II (remote) Control to Viewing Monitor				**CABLING** 50 yd drum 4-way blocks, long 4-way blocks, short Extension Adaptors			
Telecine/telejector pedestal Display bench Overhead mirror and assembly Tilt easel (with mirror) Tool kit Lens tissues Mirror cleansers				Background Roll Caption rack(s) Caption kit			

R¹ : required   L: loaded   R² : returned

Appendix

SOME TYPICAL EQUIPMENT COSTS AND APPLICATIONS
(See also key on pages 138–9)

Cassette VTR (£180–Monochrome) (£290–Colour/ Monochrome)	Records and plays back on ½-inch tape.	*A, B, C, D, E, F*
Teleplayer (£80 to £360)	Uses cartridge-loaded film, plastic tape or plastic discs to play back commercially produced software only.	*C*
Portable TV "Pack" (£600)	Works from batteries or mains to record with hand-held camera and recorder slung from shoulder.	*B, D, E*
One-inch Plumbicon Camera Tube (£370)	Can be fitted to replace cameras with Vidicon tubes to reduce noise, insensitivity, and picture lag at low light levels.	*A, B, C, D, E*
Electronic Editing Video-tape Recorders (from £1400)	Can be synchronized with a playback recorder to make clean "cuts" without picture rolls.	*A, B, C, E*

Colour Camera (£3000-£4000)	Single- and two-tube cameras, simple to operate and avoiding the high cost and complexity of the three- and four-tube colour cameras already available.	*A, B, C, D, E*
Colour Circuit Boards and Adaptors (£500-£800)	Will adapt certain video-recorders such as Ampex 7003 and IVC 601 for colour recording and playback.	*A, B, C, D, E*
Remote Control Video-tape Recorders (from £1800)	Make it possible for stop and start, fast rewind and fast forward wind to be operated from the viewer's end whilst the recorder remains in the CCTV studio or distribution centre.	*A, C, F*

KEY

A   *Relay* of e.g. conferences, discussions, announcements.

B   *Observation* of unscripted activity for e.g. time and motion study, child behaviour, athletics.

C   *Illustration*, using magnification and/or superimposition to reinforce live lecture/demonstrations. TV microscopy is particularly useful. Stimulus and exposition—short programmes to explain a process, stimulate interest, promote discussion, creative writing, or further inquiry.

D   *Self-assessment*—enabling teachers, doctors, nurses, interviewers, athletes and many others to try themselves out and progressively improve their own performances in the light of what they learn from successive replays.

E   *Project work*—in which television is used by students to record their own projects and to explore the creative possibilities of the medium.

F   *Distribution*—many different groups of viewers can be reached by the use of (*a*) a cabled network, reaching groups separated geographically, and/or (*b*) using a video-tape or many copies made from a master video-tape, which can be replayed at other times or in other places as required.

Bibliography

ARNHEIM, Rudolf   *Art and Visual Perception*   Faber 1967
ATKINSON, N. J.   *Modern Teaching Aids*   Maclaren 1966
BRENN, J. W.   *Audio Visual Instruction*   McGraw-Hill 1959
CABLE, R.   *Audio Visual Handbook*   ULP 1965
COPPEN, Helen   *Aids to Teaching and Learning*   Pergamon 1968
CORDER, P. S.   *The Visual Element in Language Teaching*
   Longmans 1967
ERIKSON, C. W.   *Fundamentals of Teaching with A V Technology*
   Macmillan 1965
GIBSON, Tony   *Experiments in Television*   EFVA 1968
GIBSON, Tony   *The Practice of ETV*   Hutchinson Educational
   1970
HEDGECOE, J., and LANGFORD, M.   *Photography: Materials
   and Methods*   Oxford Paperbacks for Artists 1971
LeFRANC, R.   *Les Techniques Audio-visuelles*   Bourrelier 1961
GAGENE, R. M.   *The conditions of Learning*   Holt, Rinehart &
   Winston
MacLEAN, R.   *Television in Education*   Methuen 1968
SAUSMAREZ, M. de   *Basic Design, The Dynamics of Visual Form*
   Studio Vista 1964
SCUPHAM, John   *The Revolution in Communication* (Technology
   and Humanities series)   Holt, Rinehart & Winston

Index